Lecture Notes in Computer Science 11415

Commenced Publication in 1973
Founding and Former Series Editors:
Gerhard Goos, Juris Hartmanis, and Jan van Leeuwen

More information about this series at http://www.springer.com/series/7407

Madalena Chaves · Manuel A. Martins (Eds.)

Molecular Logic and Computational Synthetic Biology

First International Symposium, MLCSB 2018
Santiago, Chile, December 17–18, 2018
Revised Selected Papers

 Springer

Editors
Madalena Chaves
Inria Sophia Antipolis - Méditerranée
Inria
Sophia Antipolis Cedex, France

Manuel A. Martins
Department of Mathematics
University of Aveiro
Aveiro, Portugal

ISSN 0302-9743   ISSN 1611-3349   (electronic)
Lecture Notes in Computer Science
ISBN 978-3-030-19431-4   ISBN 978-3-030-19432-1   (eBook)
https://doi.org/10.1007/978-3-030-19432-1

LNCS Sublibrary: SL1 – Theoretical Computer Science and General Issues

This Springer imprint is published by the registered company Springer Nature Switzerland AG
The registered company address is: Gewerbestrasse 11, 6330 Cham, Switzerland

Preface

It is a pleasure to introduce this volume, a collection of several papers presented at the International Symposium on Molecular Logic and Computational Synthetic Biology, which took place in Santiago de Chile, during December 17–18, 2018. Molecular logic, which focuses on computing logical operations on molecules, is a fruitful conceptual crossover between chemistry and computation. One of the goals of the symposium was to explore the potential of molecular logic frameworks to study the emerging behavioral patterns in biological networks, combining discrete, continuous, and stochastic features, and resorting both to specific or general-purpose analysis and verification techniques.

As a motivation for this symposium, the remarkable and fascinating advances in synthetic biology, which permit us to construct de novo circuits with biological components to perform specific functions. To cite only a few among many interesting examples, synthetic circuits have been constructed for autonomous oscillators in bacterial or mammalian cells (see the work of Elowitz and Leibler in *Nature* 403, 2000, or that of Tigges and co-authors in *Nature* 457, 2009), reproduce logical gates (see the work of Gander and co-authors in *Nature Communications* 8, 2017), or yet synthetically control an existent component, such as the growth switch for management of resource allocation presented by one of the keynote speakers, H. de Jong (see *PLoS Comp. Biol.* 12, 2016). Synthetic biology can also be viewed as a powerful tool for "classic" biology, since the reconstruction of a "natural" system allows for a better understanding of its mechanism and function. The engineering and implementation of these synthetic systems poses very challenging problems from a biological point of view: there are difficulties inherent to choosing suitable components and assembling them in a cellular environment; this often introduces context-dependent technical complications and raises the question of how to guarantee that each circuit part behaves as predicted, once they are connected with each other in a given environment (see the work of Del Vecchio in *Trends in Biotechnology* 33, 2015).

To overcome these challenges in the design of synthetic biology circuits, a first fundamental step is the mathematical modeling of the synthetic system, to analyze the interaction between its components and predict its dynamical behavior in different situations and in response to a variety of signals and stimuli. In accordance with the driving themes of the symposium, two common guidelines can be identified throughout all contributions to this volume: first, the modeling, analysis and understanding of *interaction networks* for biological systems; and, second, the methodology used to represent and study these biological interaction networks, based on *hybrid or logic frameworks*.

Here, a *biological system* is generally understood as a family of biological components or species, which interact to influence, regulate, or control each other's behavior. The *interaction network* represents the laws that govern the behavior of the biological system and permit us to study the time evolution of the system, its dynamics as an emergent property of the structure of interactions among the system's components. Most of the papers presented at the symposium focused on gene regulatory networks, which describe gene and protein expression in response to various forms of stimuli; other examples include chemical reaction networks and a disease spreading network.

At the methodological level there is a common search for frameworks that contain a high level of abstraction but nevertheless retain a good capacity for quantitative description and computation. This search leads to the development of methods that bridge the gap between continuous and discrete models, such as hybrid or piecewise affine models, temporal and linear logics, as well as different extensions of Boolean networks. These *logic-based frameworks* offer multiple advantages: Because they require a lower level of detail, such methods are suitable for representing large and complex systems, which often include many non-measured variables and unknown parameters. There is a large range of algorithms and computational tools (formal verification tools such as model checking; dynamic logics; graph analysis tools) that can be used to implement, interrogate, expand, and analyze the models.

This volume opens with two survey papers, for a welcome overview of the history and state of the art on the main topics of the symposium. The remaining papers in this collection are grouped into four sections, dedicated to hybrid and switching methods, Boolean models, biochemical reaction networks, and ending with specific examples of applications to gene regulatory networks.

Surveys: The first survey, by Fuentes et al., traces the history of molecular logic in biology and chemistry. In their discussion, the authors examine the pertinence of a using a molecular logic philosophy to model biological systems, based on the observation that the intracellular processes in biological organisms often appear to satisfy some "logical operations." This survey also briefly reviews four logical modeling frameworks, with a focus on Boolean models, whose dynamics are governed by a set of logical rules.

The second survey is authored by one of the symposium's keynote speakers, A. Madeira, who overviews formal verification methods based on dynamic logic. The formalisms that form the basis of dynamic logic are briefly introduced, followed by a discussion of several contributions to the systematic building of multi-valued dynamic logics. This generic method opens the way to a tailored construction of dynamical logics (targeted at synthetic biology, for instance), based on the definition of suitable atomic programs.

Hybrid and switching continuous models: This section contains two papers dealing with hybrid models that combine continuous dynamics with discrete jumps.

The work of Rocha characterizes the dynamics of a basic model for disease spread consisting of three variables, the populations of susceptible, infected, and recovered (SIR) individuals, together with a strategy for agent intervention. The rate of infection is given by a piecewise linear function of time, where switching times are determined by the agent strategy. This hybrid SIR model is more flexible and generates a larger set of solution profiles then standard SIR models.

In the second paper of this section, Huttinga et al. introduce a class of switching differential equations where the production or synthesis rates are piecewise constant and the degradation rates may satisfy certain nonlinear properties. The authors show that these new systems still admit a finite partition of the state space and the parameter space can also be decomposed into a finite number of regions. This combinatorialization of state and parameter spaces allows for the calculation of discrete state transition graphs, which represent the global dynamics of the system across all parameters.

Extensions of Boolean networks and dynamic logic: Two papers compose this section dedicated to extensions of Boolean models.

Figueiredo et al. propose to include reactive modeling into Boolean networks: this means that the edges in the directed graph can be altered as the system evolves. The authors then find a relationship between the attractors of standard Boolean networks and those of reactive Boolean networks, and also between these and the steady states of a corresponding piecewise linear model.

Goldfeder et al. develop software tools for abstract Boolean networks, a class of models where the updating rules are only partially known. In this tool, experimental observations can be encoded in linear temporal logic. Model checking tools can then be used to verify the properties and dynamics of the network motifs.

Biochemical reaction networks: In this section, Veloz et al. analyze a special class of biochemical reaction networks, called "closed reaction networks," which satisfy a set of well-defined formal properties and provides relationships between topological properties of the networks and its dynamical stability. They introduce the notion of separability to decompose a closed network into its parts and better characterize the dynamics of each part.

Two application examples: Finally, two applications to specific biological systems are presented.

A model for breast cancer progression is developed and studied by Despeyroux et al. More precisely, the dynamics of circulating tumor cells are modeled as a set of rules in linear logic, a framework which allows the authors to establish reachability properties of the model, as well as the existence of oscillations, using the Coq proof assistant.

The second paper, by Berríos et al., analyzes imaging data originating from mouse meiosis, which is a specific form of cell division. The goal is to determine the surface taken by chromatin in each image. The tools used for image analysis are a clustering process based on random chromatin neighborhoods and an association process called "P-percolation."

We would like to acknowledge two sources of support: The symposium was promoted by the project Klee - Coalgebraic Modeling and Analysis for Computational Synthetic Biology (POCI-01-0145-FEDER-030947), an R&D project supported by the Portuguese Foundation for Science and Technology; our work was also supported in part by a France–Portugal partnership PHC PESSOA 2018, Campus France #40823SD.

To conclude, we warmly thank all authors, invited speakers, members of the Program Committee, members of the local Organizing Committee, and all the participants for their work and contributions, which helped make this symposium such an attractive and successful event.

March 2019   Madalena Chaves
                                                         Manuel A. Martins

Organization

Program Chairs

Madalena Chaves   Inria Sophia Antipolis - Méditerranée, France
Manuel A. Martins   University of Aveiro, Portugal

Program Committee

Luís Barbosa   University of Minho, Portugal
Benjamin Bedregal   Federal University of Rio Grande do Norte, Brazil
Mario Benevides   Federal University of Rio de Janeiro, Brazil
Marcello Bonsangue   Leiden University, UK
Luca Cardelli   Microsoft Research, UK
Claudine Chaouiya   University of Aix-Marseille, France
Daniel Figueiredo   University of Aveiro, Portugal
Claudio Fuentes   CEAR, University of Diego Portales, Chile
Sicun Gao   University of California, USA
Monika Heiner   Brandenburg University of Technology, Germany
Hidde de Jong   Inria Grenoble - Rhône-Alpes, France
Marta Kwiatkowska   University of Oxford, UK
Alexandre Madeira   University of Minho, Portugal
Carlos Martin-Vide   University of Rovira i Virgili, Spain
Renato Neves   University of Minho, Portugal
Loïc Paulevé   CNRS/LRI University Paris-Sud, Orsay, France
Ion Petre   Abo Akademi, University of Turku, Finland
Tatjana Petrov   University of Konstanz, Germany
Élisabeth Remy   IML, University of Aix-Marseille, France
Eugénio Rocha   University of Aveiro, Portugal
Marie-France Sagot   Inria Grenoble - Rhône-Alpes, France
Regivan Santiago   Federal University of Rio Grande do Norte, Brazil
Amilra P. de Silva   Queen's University Belfast, UK
Ana Sokolova   University of Salzburg, Austria
Meng Sun   University of Peking, China
Carolynn Talcott   SRI International, USA
Antonio
    J. Tallón-Ballesteros   University of Seville, Spain
Delfim F. M. Torres   University of Aveiro, Portugal
Tomas Veloz   IFICC, Chile
Boyan Yordanov   Microsoft Research, UK
Paolo Zuliani   Newcastle University, UK

Organizing Committee

Daniel Figueiredo	University of Aveiro, Portugal
Claudio Fuentes	CEAR, University of Diego Portales, Chile
Pablo Razeto-Barry	IFICC, Chile
Tomas Veloz	IFICC, Chile

Contents

Contents

Molecular Logic: Brief Introduction and Some Philosophical Considerations

Claudio Fuentes Bravo[1]([⊠]) and Patricio Fuentes Bravo[2]

[1] CEAR – Centre for the Study of Argumentation and Reasoning
of the Faculty of Psychology at Diego Portales University, Santiago, Chile
claudio.fuentes@udp.cl
[2] Department of Experimental Oncology Laboratory of Stem Cell Epigenetics,
European Institute of Oncology, Milan, Italy

Abstract. In the present article a brief historical and systematic introduction to the field of molecular logic are proposed. Some relevant philosophical consequences derived from the technical treatment of this topic are also exposed. These consequences are made explicit in three fundamental questions. Some of the proposed methods for the representation of the intracellular molecular dynamics are also presented and the advantages and limitations that the different methods exhibit when modeling natural biological circuits are evaluated. The Boolean approach to molecular logic is considered with special attention in this article, emphasizing that "logic gates" have proven to be functionally appropriate for analyzing experimental information, however, they present limitations to capture complex biological processes. In relation to this last point, the problem presented by the modeling of continuous variables through discrete systems is studied in depth. It is explained then the need to have adequate logic to the phenomenon and its characteristics.

Keywords: Molecular logic · Boolean approach · Logic gates · Philosophy

1 Introduction

The experimental study of the digital features of a cellular organism like the bacteria *Eschericcia coli* (*E. coli*) confirmed the existence of properties and biochemical-molecular principles which are present in every diverse living forms on Earth, granting the proper context to develop a logical and predictive comprehension for the strategies of Life [13].

More than 40 years ago, Lehninger called "The molecular logic of living state" to these "biochemical-molecular generalizations" which are present in every living organisms on Earth [30]. Indeed, the first general reference to an "intrinsic molecular logic" is found by 1993 in the paper [40] from de Silva, et al. In this paper, the work of the authors mainly consisted in the implementation of a system with a luminescent signaler, called *Photoinduced Electron Transfer* (PET,

© Springer Nature Switzerland AG 2019
M. Chaves and M. A. Martins (Eds.): MLCSB 2018, LNCS 11415, pp. 1–17, 2019.
https://doi.org/10.1007/978-3-030-19432-1_1

for short). This technique has then allowed a posterior development in the experimental approaches using logic gates.

In the same paper, de Silva et al. define the Molecular Logic-based Computation, as an approach that is applied to the molecules and chemical systems which have an innate capacity to compute, even if in a rudimentar way, like machines based in transistors, semiconductors or people. This conception of molecular logic is certainly close to the functional ideas of the *Natural Computing*, which we can relate to the famous quote of R. Feynman in 1960: "There's Plenty of Room at the Bottom".

According to Jiménez and Caparrini (see [17]), the physical limitation in the computational speed of a conventional sorter and the first resolution of a computationally hard mathematical problem by manipulating DNA molecules in the laboratory imply a fundamental scientifically advance in the field of molecular computation: the definitive step from a theoretical perspective (*in info*) to an experimental perspective (*in vitro*).

Thus, in the topic of molecular logic we can gather the interest of different fields which can be summarized in two main classes of scientific groups: on one hand, the biochemical and, on the other, mathematical and philosophic.

We can see the latest in the following way: the unveil/modeling of the logic underlying to the processes concerning the intracellular dynamics is a problem associated to the interest of biologist and chemists. Moreover, from the point of view of science philosophy, it can be evaluated as contributing to the mechanistic-like paradigm of molecular logic, a problem within the philosophic interest. Related to this, within science philosophy, a successful explanation in molecular biology would require the identification and manipulation of variables in a casual mechanism. In other words, this would require the understanding about how the diverse variables interfering in certain mechanism act and interact to produce the phenomenon. Indeed, the "natural genetic networks", a relatively recent scientific construction in Biology, can be useful to obtain a mechanistic explanation to the biological processes taking the following three fundamental properties as starting point: (1) the considered network is complex, (2) the components of the network interact with each other and (3) we can attribute specific functions to each component.

1.1   A Starting Point

Regarding the philosophy of Biology and starting from the decade of 1960, more and more doubts arose about the existence of universal laws, *i.e.* rules that could be considered as necessary universal generalizations. As example of this conceptualization we can recall the relevance acquired by the concept of *earthboundedness* [42]. Nevertheless, there is a relative consensus regarding at least two general facts for the understanding of life as a scientific phenomenon

- The life on Earth is genetically connected by a evolutive past which has been occurring for about four billion years.
- The organic compounds composed all organisms have been selected during evolution through adaptive processes that occurred continuously while performing specific biochemical or cellular functions.

Taking this into account, we can ask some questions related to the development of a logic of intracellular dynamics with great philosophical repercussion:

- Do biological organisms on this planet exhibit a "common logic" at the level of their intracellular processes?
- Can the "logical operations" we perceive in biological organisms become an instrument for naturalizing our concept of logic?

2   Antecedents

2.1   The Concept of Molecular Logic in the Biological Sciences

As we know, the structure of organic compounds obeys the physical and chemical laws that describe the behavior of inanimate matter. In an apparent paradox, we realize that it is the interaction of these inanimate organic compounds itself that maintains and perpetuates animate life. This is an apparent paradox because from the past it was believed that to the inanimate matter could not be by itself anything else than its own inanimate nature. This metaphysical contradiction seems to fade after the verification of intracellular dynamics. The structures, mechanisms, processes and biochemical adaptations within cells, the basic functional unit of an organism, are shared by plants, animals and even unicellular organisms in a fundamental chemical pattern: the DNA molecule.

The DNA, at its level of primary structural organization, is a linear sequence of molecular elements called nucleotides (or bases) which constitute what is known as the genome of living organisms. From another perspective, we can consider a gene as the elementary structure of a genomic sequence since it occupies a specific position in a chromosome and determines the expression of a protein in an organism. With the development of molecular biology we have learned that DNA can act as a digital storage device (this analogy is fundamental to the statements that follow) which, in its turn, can be read, copied and replicated during events such as cell division or even experimentally, using appropriate reagents and molecules [6, 47].

We can note that more than four decades have passed since Lehninger's work and the advent of functional genomics has made possible the characterization of the molecular constituents of life. On one other hand, at the technical level, "conserved processes" allows us today to use models for organisms and to explore and infer the function of human genes. On the other its helps us to situate the same genes in the normal and pathological context. Thus, the field of study that we know as genomic research, has drastically reduced the gap of ignorance about biological systems in few years, driven the vision that these systems are composed basically of two types of information [21]:

– Genes, which encode the molecular tools which carry out the functions of life
– Genetic network, which specify how genes interact and are expressed.

From other point of view, the integration of different organizational levels supports the idea that cellular functions are distributed among groups of heterogeneous components that interact with each other within larger networks. Thence, the organization of the proteome [37] as a network composed of interacting proteins and other secondary components which are inter-converted due to an intricate metabolic network that has been called the metabolome in [31]. However, in the context of the referred paper, is suggested that the structure of these networks is ruled by what we can call a "logic" and this logic itself is present in the entire and more complex organism at a macroscopic level.

2.2   The Role of Systems Biology

Systems biology has been the disciplinary field that has tried to gather information in a coherent way from each of these different levels of analysis, referring in turn to various individual biological processes. Within this field, many researchers work to generate integrated (computational) models of systems in which cellular components and networks interact within temporal, spatial, and dynamic physiological contexts from the available data [8, 15]. In this context, we can affirm, in its turn, that the structural organization of a natural or biological system refers fundamentally to the material configuration that relates the components of the system.

For instance:

– The structure of the genomes plays a fundamental role concerning the way in which the information contained in the genes is connected to the phenotypic expression pattern. At the same time,
– the genotype is determined by the information contained in the DNA sequence.
– the phenotype is determined by the context-dependent expression of the genome, and
– the genetic networks which interpret the context and orchestrate patterns of expression.

Genetic networks, exhibit a feature of great utility for a logical understanding of gene expression, since they can be described as circuits of interconnected functional modules (IFM). Each IFM is composed of specialized interactions between proteins, DNA, RNA, and small molecules. In this context, a module corresponds to the simplest element of a regulatory genetic network (RGN). This RGN is composed of a promoter, the genes expressed from that promoter, and the regulated proteins (as well as their binding locations in the DNA) that affect the expression of that gene.

The described IMF modules can be considered "logical devices" *i.e.*, devices that "behave similarly" or "can be represented as" electronic devices with a Boolean function where the output (product) is the level of expression of a gene. This level of expression is determined by the amount of mRNA or protein produced, and the input (substrates) are the factors that affect the expression of a particular gene (regulatory proteins, transcription factors, etc.).

This structure with functional independence between multiple subunits of genomes, eases the modular nature at higher organizational levels [12]. This plasticity (capacity of living organisms to adapt) can be considered as the result of the action of complex systems composed of components that interact with each other. However, this makes their behavior difficult to predict and prone to unexpected results. That is, the plasticity exhibited by living organisms seems to complicate the possibility of being captured by mechanical or algorithmic procedures.

2.3 Lac. Operon

Within a cell occurs an extraordinary and incessant flow of information (expressed by multiple and complex specific actions of interaction). The importance of cellular information exchange is particularly recognized when one studies how it flows from genes to proteins.

An example of this is the regulatory mechanism present in prokaryotic (bacterial) cells, where multiple genes, involved in a process of metabolism of a certain carbon source, are expressed in a coordinated manner with a single promoter, in a natural genetic logic module called an operon. The lac operon of the bacterium Escherichia coli (*E. coli*), is one of the best characterized prokaryotic systems of gene regulation. It is composed of three genes – lacZ, lacY, and lacA – involved in the metabolism and binding of the disaccharide lactose. This mechanism can be described in terms of molecular interactions between DNA, proteins and metabolites (intermediates and products of metabolism). Indeed, these interactions make sense when they are modeled by negative feedback loops that process information about the presence of lactose on the environment to regulate the rate of transcription of the lac operon [16].

Then, the relevance of determining the components of natural genetic networks such as the lac operon is that they become ideal candidates for the construction of artificial genetic networks [18]. In order to comprehend a natural logic circuit that operate within a cells, such as the lac operon, it must be separated into the individual processing elements or logical modules carrying specific information (functions). This separation depends or is made possible through chemical isolation, which can be obtained from spatial localization (compartmentalization) or/and chemical specificity. This means that different information can be stored in different places and a wide variety of connections between logical modules can be formed and reformed through the diffusion of chemical agents (reagents).

An example of the above is a signaling system such as the one of chemotaxis in bacteria which is an extended module that achieves its isolation due to the binding specificity of the initial chemical signal (chemoattractants) to a receptor protein, and due to the interaction between signaling proteins within the cell.

Thus, at this point, it is necessary to try to characterize how a logical module operates.

2.4 Natural Logic Modules

A good example of a natural logic module is the one that controls the precise distribution of chromosomes in the resulting cells during mitosis. In its turn, it contains modules that assemble the mitotic spindle:

- A module that controls the alignment of the chromosomes in the spindle;
- A cell cycle oscillator which regulates the transactions between interphase and mitosis.

The way these modules act depends on how the various components are connected to each other and is affected by the shapes of the response curves that determine the kinds of those interactions. On the other hand, the identification of the logical modules used in cellular systems must generate an inventory to define the logical tools that are available within the cells [12].

This shows that the representation of a natural logic circuit is useful primarily to establish and understand the following three features of the system:

- How do logical modules constitute a circuit which processes informations within a cell;
- What are the logical relations between components;
- How do information flows inside the circuit.

3 Modeling Methodologies

During the last century, multiple modeling methodologies have been developed for biological circuits. Among them it is possible to count mathematical models based on differential equations, Boolean probabilistic networks, Petri nets, Bayesian networks and many others. During this section we recall some of them, highlighting the advantages and the disadvantages.

3.1 Models with Differential Equations

Mathematical models are generally based on ODEs (ordinary differential equations) and PDEs (partial differential equations) that require the a priori knowledge about the interaction pattern between the analyzed components. Moreover, the result of the model strongly depends on the initial concentration and kinetic constants of the same components (see Fig. 1).

A disadvantage presented by these models is that they become more and more difficult to obtain and analyze whenever the number of interdependent variables increase and the relations between them depend on qualitative events (a concentration threshold, for example) [34].

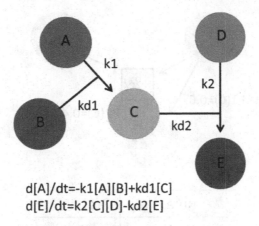

d[A]/dt=-k1[A][B]+kd1[C]
d[E]/dt=k2[C][D]-kd2[E]

Fig. 1. Differential-equations models using mass-action kinetics depic regulatory mechanisms by defining rates of change in network species concentrations.

3.2 Probabilistic Boolean Networks

In the Boolean networks introduced by Kauffman [20], gene expression is constrained to only two levels: ON and OFF. The expression level (state) of each gene is functionally related to the expression states of some other genes, using logical rules. However, as a result of scarce experimental data or incomplete comprehension of a system, several candidate regulatory functions may be possible for an entity. This entails the search to express uncertainty in the regulatory logic. The Probabilistic Boolean networks (PBN) introduced by Shmulevich et al. [39] extend the classical Boolean network. In such model, an entity can have several regulatory functions, to each of which is given a probability based on its compatibility with previous data. At each time step, each variable is subjected to a regulation function that is randomly selected according to the defined probabilities (Fig. 2). The model is stochastic with an initial global state that can lead to many pathways harboring distinct probabilities. Thus the new model, the PBN, gives rise a sequence of global states that constitutes a Markov chain [14].

3.3 Petri Nets

Petri nets (PN) [32] are mathematical models which can be conceived as a generalization of automata and allows to express a system with concurrent events. A PN is a graph with two kinds of nodes: places and transactions. The places represent the resources of the system while the transitions correspond to events that can change the state of the resources. The oriented edges (*weighted arcs*) connect places to transitions and *vice versa*, representing the relations between resources and events. The state of the system is represented using symbols called *tokens* embedded on places; a place can contain multiple tokens (Fig. 3). On its turn, distinct assignments of tokens over places induce different states in the system. To each one of these assignments, we call a *marking* [7,43].

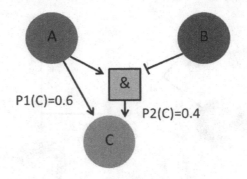

P1(C)=0.6

P2(C)=0.4

Rules	Probability
C=A	0.6
C=A AND NOT B	0.4

Fig. 2. Basic building blocks of the PBN. The model consists of 3 nodes with one activation edge and one partial inhibition edge. The weights of both edges are expressed as selection probability next to the arrow (upper). Two representative Boolean rules were assigned with the corresponding selection probabilities (Pj(i)) to represent the example model in PBN format. The truth table of the model evidences the state values according to different inputs. Once both inputs (A and B) are active, the output (C) has a probability of being ON at 0.6 and of being OFF at 0.4 according to the selection probability of Boolean rules (lower).

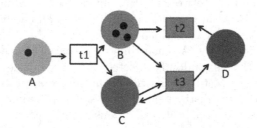

Fig. 3. The classical Petri net model is a network composed of places (circles) and transitions (rectangles). The connectors, called arcs are oriented to, and between, the places and transitions. The tokens (black circles) are the dynamic objects. Finally the state of the petri net, the marking, is determined by the distribution of tokens over places. Therefore, the initial marking is 1, 2, 0, 0.

A relevant technical advantage of these structures is that the most of problems are decidables in PNs. This decidability can be obtained *via* a Karp-Miller tree [19]. Moreover, it is known that the reachability problem is decidable at most in exponential time. Still, it is also known that a Petri net is able to model a system with parallel evolution or concurrent events composed of several processes which cooperate to attain a common goal.

3.4 Bayesian Networks

With the advent of the next-generation sequencing (NGS) revolution [22, 29] the use of Bayesian networks has emerged as a very promising method. They are becoming increasingly important in the biological sciences for inferring cellular networks, modelling protein signalling pathways, systems biology, data integration, classification, and genetic data analysis.

Bayesian networks, or alternatively graphical models, are very useful tools for dealing not only with uncertainty, but also with complexity and (even more importantly) causality — — — — — — — — — — — — — — — — — — —. These provide an accurate and compact representation joint probability distributions (JPDs) and for inferring from incomplete data and to adapt the number of parameters to the size of the sample. Bayesian networks are particularly useful to describe processes of components which interact locally, i.e., processes where the values of each component variable directly depend on the values of a relatively small set of component variables (Fig. 4).

These models are acyclic digraphs whose nodes represent random variables in a Bayesian sense: they can be a observable concentration, unknown parameters,

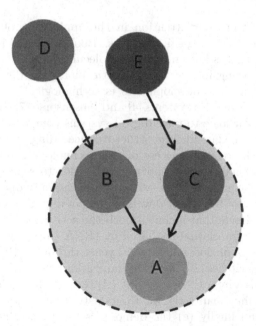

Fig. 4. Gene regulatory networks provide a natural example for BN application. Genes correspond to nodes in the network, and regulatory relationships between genes are shown by directed edges. In the simple example above, A is conditionally independent of D and E given B and C. $p(A, B, C, D, E) = p(D)p(E)p(B—D)p(C—E)p(A—B,C)$.

latent variables or hypothesis. It is relevant to mention that similar ideas the ones mentioned in BN and which are currently being used, can be applied to simple (non-directed) graphs and, possibly, even to cyclic graphs, like Markov networks (MN). In Bayesian networks, the expression of each gene is represented by a JPD variable, which describe how genes regulate each other. The graphic representation shows clearly the places where regulatory relationships exist between genes. In a BR, the tips (ends) have to form a directed acyclic graph (DAG) which is a graph without cyclic paths (including loops). In this way, JPDs are understood in a compact manner, reducing the size of the model when exploring conditional independence relations (two variables are independent given the state of a third variable).

BR can also be interpreted as a casual model which generates data. Then, the arrows (directed tips) in the DAG can describe relations or causal dependences between variables [29, 38]. However, it is known that Bayesian networks present some limitations. The modeling of genetic networks require that these models must be represented as acyclic graphs and, therefore, they are not able to to represent feedback control mechanisms. To accomplish this, *Dynamic Bayesian Networks* have been developed (DBN). These are networks which are able to describe temporal processes like feedback loops [49].

4   Boolean Molecular Logic

In this section, we focus our attention in the application of boolean logic to represent some specific biological networks. Boolean logics have been widely applied in the field that is known itself as molecular logic.

A subject of growing interest in synthetic biology is digital gene circuits. The main interest in these developments is technological, *i.e.*, their potential constitution as bioinformatics systems [4] and biosensors [27]. In this context, we call boolean molecular logic to refer digital systems composed of Boolean gates which have been designed in different structures according to diverse biochemical mechanisms [26]. The relative success of this technology can be related to the fact that transcriptional controls showed to be able to reproduce, in bacteria, the majority of Boolean gates with two inputs using only one or two regulated promoters [41]. Some of the recent developments in digital Boolean systems are: (a) a first complex design that included both the activation of the promoter and the regulation of the translation mediated by tRNA, applying a AND gate in *E. coli* [2]; (b) the design of Boolean gates in yeast through mRNA structures like ribozymes and riboswitches [23, 46]; the organization of mammalian cell in digital circuits based on RNA interference [36, 48]; (d) the translation of regulation PBN models, used in mammalian cells, for the construction of logic gates based on a single cell [3]; and finally (e) the demonstration that cellular coupling can be a solution for the modular design of the logic circuits in yeast [35] and the construction of a NOR gate in *E. coli* by Tamsir, Tabor & Voigt [44].

4.1   Constructing Molecular Boolean Gates

Let us analyze in more detail the application of Tamsir, Tabor & Voigt. They developed a method which allows to compute Boolean logical operations within cells in an analogous way to the artificial integrated circuits, where the logic gate is coded in DNA and operates along with other biomolecules. Specifically, the authors tried to show that biochemical processes such as transcription of a repressor in a specific cell, can be seen as a Boolean logic gate from a functional point of view. Bearing that goal in mind, Tamsir and his collaborators built a collection of logic gates using a colony of *E. coli* bacteria by means of the functional complete NOR gate.

A NOR gate is a negates the output of a OR gate, which is a disjunction between both inputs. The input values of logic gates are binary and classically represent truth values with 1 being "true" and 0 being "false". In practice, the 0,1 inputs can also be conceived as absence and presence of a certain molecular component or output. The first experiment performed by Tamsir and his colleagues consists into obtaining four different responses of a *E. coli* bacteria. These responses depend on the combination of presence (1) and absence (0) of two inputs components – arabinose (Ara) and anhydrotetracycline (aTc) – which are molecules that promote the genetic transcription within the cell. This components activate distinct and independent promoters – Pbad and Ptet, respectively. Once active, these promoters activate a gen repressor called CI which, in its turn, suppress a gen called YPF that allows the production of a fluorescent protein, when active. This fluorescence is considered the Boolean output variable of the system. In this way, the presence of ArA or aTc activates CI, which inhibits the production of YPF (therefore, the output is 0). However, when neither Ara nor aTc is present, CI is not active and YPF is produced (therefore, the output is 1).

After obtaining a NOR gate, Tamsir and his collaborators built a XOR gate from the combination of three NOR gates. Each NOR gate was programmed within an independent *E. coli* bacteria. A XOR gate accepts two inputs and consists of a gate whose output is 1 if exactly one inputs is present. In any other case, its output is 0. Thus, if exactly one of the input molecules are present, the result is that YPF is produced. Conversely, if this is not the case, there no production of YPF.

In the Fig. 5 of the paper [44], we can analyze the behavior of one of the cells that make up the XOR gate (cell 1, Fig. 5). In that cell we can see that when there is no presence of any of the inducing molecules, the cell has high fluorescence values (10^3 AU). Differently, when both substances are present, the fluorescent substance only reaches a value close to 10^1. Also, in another cell (cell 4, Fig. 5) functioning as an output of the XOR gate, it can be seen that the fluorescence reached is lower when both or none of the inductors is present with respect to the case in which either of the two is independently present.

The problem which generally can be pointed to Boolean network and that can be extended to the proposal of Tamsir and his colleagues is the inherent determinism. It does not seem a good solution, in order to obtain a greater formal robustness, to avoid the stochasticity of gene expression or the

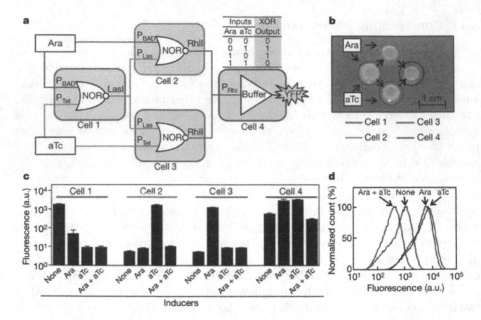

Fig. 5. Construction of an XOR gate by programming communication between colonies on a plate. A, Four colonies each composed of a strain containing a single gate are arranged such that the computation progresses from left to right, with the result of each layer communicated by means of quorum signals. The inputs (Ara and aTc) are added uniformly to the plate. B, Spatial arrangement of the colonies. C, Each colony responds appropriately to the combinations of input signals. Fluorescence values and their error bars are calculated as mean s.d. from three experiments. D, Cytometry data for the XOR gate (cell 4). [Tamsir A, Tabor JJ, Voigt CA. Nature. 2010;469(7329):212-5].

uncertainty associated with the measurement of its processes, which can occur due to experimental noise and the latency of interacting variables, in the modeling. The inference of a single deterministic function will inevitably result in a poor predictive accuracy and insufficient knowledge of the architecture of the network. This occurs since it is a small sample relative to the number of genes. That is, it must be taken into account that the stochastic effect will inevitably spread from the micro level to the macro level [1].

4.2   An Alternative to [44]

In [35], circuits with better digital responses were implemented. In short, the claim that the digital approach is not the best option to describe experimental behavior can be reformulated to a new one suggesting that, probably, the colonies used in [44]. They are not the best to implement a digital circuit.

Using computer simulations, Marchisio and Stelling, [50] showed that, in order to improve the separation of gate signal, the multiple integration of the final gate (*i.e.*, the transcription of unity of coding for the circuit output) could be a valid alternative to the reengineering of the final gate promoter. With this strategy they were able to construct the only two gates that did not work in their original design. Lastly, the distributed output architecture was applied [35] to genetic networks. This design strategy showed clear advantages with respect to the final gate architecture: less wiring between the gates, fewer genes in the network, and the output are levels easy to predict.

4.3   Biological Detail of the Essay

We argued that the kind of analysis in [44] is not the best since a Boolean gate is not suitable to describe the behaviour of *E. coli* colonies which take part into the creation of logical circuits. Regot et al. implement circuits with better digital results, using other colonies. We note that *E. coli* colonies would be able to apply the NOR gate and, actually, any other gate could be obtained *via* a suitable spacial distribution of these colonies.

The results obtained by Tamsir and his colleagues had a low digital quality when complex circuits where implemented. Whether in genetics or in electrical circuits, in the real world they are analogical signals, and Boolean systems are discrete. The main argument for considering that a system has a digital answer is the possibility of establishing thresholds that allow to define the low level (logical 0) and the high level (logical 1), and there are no signals between both levels. However, these thresholds must be fixed for every component in the circuit (or, in this case, colonies). These restriction is not fully satisfied, at all, in the paper of Tamsir.

Although the results in his paper are not a good example of digital behavior, the goal of Tamsir and his colleagues was to create a general method to implement digital circuits based on spatial distribution of colonies of *E. coli*. Therefore, the description of the circuit, based on NOR logic gates, was not a model that would be used to describe the dynamic of the biological system. Since the goal was to create colonies behaving as similar to logic gates as possible, we cannot infer about nothing about the a generic biological system in what relates to its digital behavior. The simulation performed by Marchisio and Stelling [50] seem to be promising to solve critical aspects of boolean systems like the codification of unity of transcription to the end of the circuit. In that sense, the application of a distributed gate architecture system to genetic networks could be a valid alternative to the reengineering of the final gate promoter, improving the efficiency and predictability of circuit.

5   Conclusion and Further Work

The thesis we support in this brief essay is that the dynamics underlying the intracellular molecular processes (molecular logic) is one and it is generalizable to all living organisms and their biological processes at different levels of complexity. This is corroborated by the fact that it is generally accepted as a scientific fact that all life on earth is genetically related by an evolutionary past and that organic compounds forming every organisms also respond to adaptive processes, which are the result, in their turn, of the performance of specific biochemical or cellular functions.

Moreover, it is necessary to refer that we have laboratory techniques, such as those described by de Silva [9], among others, that allow us to understand the dynamics of intracellular molecular processes to described them in a logic containing the essential chemical-molecular properties of organic life. The question arising as consequence would be: "What logic would be the most appropriate to represent the complexity of the processes described?". Because of this, we have reviewed several methodological approaches used to model systems describing molecular dynamics and considered its advantages and disadvantages. Here we must note that, although logic gates are functionally suitable to analyze experimental information of cellular processes, the model based on Boolean logic have both conceptual and predictive limitations when used to analyze more complex processes. The usage of distinct organizational level promote the thesis that cellular functions are distributed in groups of heterogeneous components that interact with each other within wider networks. On the other hand, many important biological variables (as concentration values and fluorescence levels) can be presented with discrete values but, strictly speaking, still are continuous variables.

At this point, we point out that we consider important to explore the technical capacity that we would eventually have to characterize this logic underlying every organisms living on Earth. John Wood talks about "heavy machinery" to refer formal theories and tools which are extremely expressive, such as dynamic logic, and their application to describe cognitive processes in Humans. However, two good paradigms to take part in the complex intrinsic molecular logic are:

- Reconfigurable systems [24] which would allow us to use, at different times, distinct logic such as multivaluated and modal logics, as well as algebraic approaches to reduce the resulting technical and conceptual gap, that must be solved in order to use, often, natural modules to redesign synthetic, reliable and robust organisms.
- Differential dynamic logic [10,33]. These systems include both continuous and discrete dynamics and are known as *hybrid*. The differential dynamic logic is a recently developed logic to work with such systems. In practice, it is a dynamic logic (see [11]), that is itself inside the class of modal logics [28], but embedding a first order structure.

The capacity that this recent and powerful mathematical developments provide to catch the underlying molecular logic is a challenge that certainly will need further work. However, we can question the following: "The choice for a logic strong enough to capture the an underlying molecular logic contributes to the understanding of the term *universal logic*, as proposed in [5]?".

The path to achieve the naturalization of logic assumes that the generalization of a molecular logic is equivalent to assuming that human cognitive processes must guide us in the achievement of a materially founded logic. This sense of naturalization of logic is broader than the one proposed in the works of Magnani (see [25]), referring exclusively to specif processes of human reasoning. The existence of a logic embedded in the processes of living organisms serves as evidence to present a different path of naturalization. The logic would be no more a mere formal construction, with no cognitive meaning, because cognitive processes are not solely performed by humans. This assumption takes us away from naturalization as conceived by Magnani. The coupling of each organism with the environment is itself a result of natural selection [45]. Then, has referred, the life on Earth seems to have an intrinsic logic.

This can lead us to think that it seems strange or, at least, a justification is needed to argue that this natural logic would only result from human cognitive mechanisms, because the same cognitive mechanisms that would realize certain mathematical principles, are embedded in organic structures in which the same mathematical principles seem to underlie. This is another challenge to study in future researches.

References

1. Acar, M., Mettetal, J.T., Van Oudenaarden, A.: Stochastic switching as a survival strategy in fluctuating environments. Nat. Genet. **40**(4), 471 (2008)
2. Anderson, J.C., Voigt, C.A., Arkin, A.P.: Environmental signal integration by a modular and gate. Mol. Syst. Biol. **3**(1), 133 (2007)
3. Ausländer, S., Ausländer, D., Müller, M., Wieland, M., Fussenegger, M.: Programmable single-cell mammalian biocomputers. Nature **487**(7405), 123 (2012)
4. Benenson, Y.: Biomolecular computing systems: principles, progress and potential. Nat. Rev. Genet. **13**(7), 455 (2012)
5. Beziau, J.Y.: Logica universalis: towards a general theory of logic. Springer (2007). https://doi.org/10.1007/978-3-7643-8354-1
6. Carell, T.: Molecular computing: DNA as a logic operator. Nature **469**(7328), 45 (2011)
7. Chaouiya, C.: Petri net modelling of biological networks. Briefings Bioinf. **8**(4), 210–219 (2007)
8. Chuang, H.Y., Hofree, M., Ideker, T.: A decade of systems biology. Ann. Rev. Cell Dev. Biol. **26**, 721–744 (2010)
9. De Silva, A.P.: Molecular Logic-based Computation. Royal Society of Chemistry, Cambridge (2016)
10. Figueiredo, D.: Differential dynamic logic and applications. Master's thesis, University of Aveiro (2015)
11. Harel, D., Kozen, D., Tiuryn, J.: Dynamic logic. In: Gabbay, D.M., Guenthner, F. (eds.) Handbook of Philosophical Logic, pp. 99–217. Springer, Dordrecht (2001). https://doi.org/10.1007/978-94-017-0456-4_2
12. Hartwell, L.H., Hopfield, J.J., Leibler, S., Murray, A.W.: From molecular to modular cell biology. Nature **402**(6761 suppl.), C47 (1999)
13. Hood, L.: Systems biology: integrating technology, biology, and computation. Mech. Ageing Dev. **124**(1), 9–16 (2003)
14. Husic, B.E., Pande, V.S.: Markov state models: from an art to a science. J. Am. Chem. Soc. **140**(7), 2386–2396 (2018)

15. Ideker, T., Galitski, T., Hood, L.: A new approach to decoding life: systems biology. Ann. Rev. Genomics Hum. Genet. **2**(1), 343–372 (2001)
16. Jacob, F., Monod, J.: Genetic regulatory mechanisms in the synthesis of proteins. J. Mol. Biol. **3**(3), 318–356 (1961)
17. Jiménez, M.d.J.P., Caparrini, F.S.: Máquinas moleculares basadas en ADN. No. 2, Universidad de Sevilla (2003)
18. Kaern, M., Blake, W.J., Collins, J.J.: The engineering of gene regulatory networks. Ann. Rev. Biomed. Eng. **5**(1), 179–206 (2003)
19. Karp, R.M., Miller, R.E., Rosenberg, A.L.: Rapid identification of repeated patterns in strings, trees and arrays. In: Proceedings of the Fourth Annual ACM Symposium on Theory of Computing, pp. 125–136. ACM (1972)
20. Kauffman, S.A.: Metabolic stability and epigenesis in randomly constructed genetic nets. J. Theor. Biol. **22**(3), 437–467 (1969)
21. Kitano, H.: Systems biology: a brief overview. Science **295**(5560), 1662–1664 (2002)
22. Koboldt, D.C., Steinberg, K.M., Larson, D.E., Wilson, R.K., Mardis, E.R.: The next-generation sequencing revolution and its impact on genomics. Cell **155**(1), 27–38 (2013)
23. Kötter, P., Weigand, J.E., Meyer, B., Entian, K.D., Suess, B.: A fast and efficient translational control system for conditional expression of yeast genes. Nucleic Acids Res. **37**(18), e120–e120 (2009)
24. Madeira, A.L.C.: Foundations and techniques for software reconfigurability (2013)
25. Magnani, L.: Naturalizing logic: errors of reasoning vindicated: logic reapproaches cognitive science. J. Appl. Logic **13**(1), 13–36 (2015)
26. Marchisio, M.A.: In silico design and in vivo implementation of yeast gene boolean gates. J. Biol. Eng. **8**(1), 6 (2014)
27. Marchisio, M.A., Rudolf, F.: Synthetic biosensing systems. Int. J. Biochem. Cell Biol. **43**(3), 310–319 (2011)
28. Marx, M., Venema, Y.: Local variations on a loose theme: modal logic and decidability. In: Finite Model Theory and Its Applications, pp. 371–429. Springer, Berlin (2007). https://doi.org/10.1007/3-540-68804-8_7
29. Needham, C.J., Bradford, J.R., Bulpitt, A.J., Westhead, D.R.: A primer on learning in bayesian networks for computational biology. PLoS Comput. Biol. **3**(8), e129 (2007)
30. Nelson, D.L., Lehninger, A.L., Cox, M.M.: Lehninger Principles of Biochemistry. Macmillan (2008)
31. Ng, A., Bursteinas, B., Gao, Q., Mollison, E., Zvelebil, M.: Resources for integrative systems biology: from data through databases to networks and dynamic system models. Briefings Bioinf. **7**(4), 318–330 (2006)
32. Petri, C.A.: Kommunikation mit automaten (1962)
33. Platzer, A.: Logical Analysis of Hybrid Systems: Proving Theorems For Complex Dynamics. Springer, Berlin (2010). https://doi.org/10.1007/978-3-642-14509-4
34. Raj, A., van Oudenaarden, A.: Nature, nurture, or chance: stochastic gene expression and its consequences. Cell **135**(2), 216–226 (2008)
35. Regot, S., et al.: Distributed biological computation with multicellular engineered networks. Nature **469**(7329), 207 (2011)
36. Rinaudo, K., Bleris, L., Maddamsetti, R., Subramanian, S., Weiss, R., Benenson, Y.: A universal RNAi-based logic evaluator that operates in mammalian cells. Nat. Biotechnol. **25**(7), 795 (2007)
37. Rual, J.F.: Towards a proteome-scale map of the human protein-protein interaction network. Nature **437**(7062), 1173 (2005)

38. Sachs, K., Gifford, D., Jaakkola, T., Sorger, P., Lauffenburger, D.A.: Bayesian network approach to cell signaling pathway modeling. Sci. STKE **2002**(148), pe38 (2002)
39. Shmulevich, I., Dougherty, E.R., Kim, S., Zhang, W.: Probabilistic boolean networks: a rule-based uncertainty model for gene regulatory networks. Bioinformatics **18**(2), 261–274 (2002)
40. de Silva, P.A., Gunaratne, N.H., McCoy, C.P.: A molecular photoionic and gate based on fluorescent signalling. Nature **364**(6432), 42 (1993)
41. Silva-Rocha, R., de Lorenzo, V.: Mining logic gates in prokaryotic transcriptional regulation networks. FEBS Lett. **582**(8), 1237–1244 (2008)
42. Smart, J.C.: 1963, Philosophy and Scientific Realism (1970)
43. Steggles, L.J., Banks, R., Shaw, O., Wipat, A.: Qualitatively modelling and-analysing genetic regulatory networks: a petri net approach. Bioinformatics **23**(3), 336–343 (2006)
44. Tamsir, A., Tabor, J.J., Voigt, C.A.: Robust multicellular computing using genetically encoded NOR gates and chemical wires. Nature **469**(7329), 212 (2011)
45. Varela, F., Thompson, E., Rosch, E.: The Embodied Mind: Cognitive Science and Human Experience. MIT Press, Cambridge (1991)
46. Win, M.N., Smolke, C.D.: Higher-order cellular information processing with synthetic RNA devices. Science **322**(5900), 456–460 (2008)
47. Woodcock, C.L.: Chromatin architecture. Curr. Opin. Struct. Biol. **16**(2), 213–220 (2006)
48. Xie, Z., Wroblewska, L., Prochazka, L., Weiss, R., Benenson, Y.: Multi-input rnai-based logic circuit for identification of specific cancer cells. Science **333**(6047), 1307–1311 (2011)
49. Zou, M., Conzen, S.D.: A new dynamic bayesian network (DBN) approach for identifying gene regulatory networks from time course microarray data. Bioinformatics **21**(1), 71–79 (2004)
50. Marchisio, M.A., Stelling, J.: Computational design of synthetic gene circuits with composable parts. Bioinformatics **24**(17), 1903–1910 (2008)

Verification for Everyone?
An Overview of Dynamic Logic

Alexandre Madeira[1,2(✉)]

[1] CIDMA, University of Aveiro, Aveiro, Portugal
madeira@ua.pt
[2] QuantaLab INESC TEC, University of Minho, Braga, Portugal

Abstract. This note, reporting the homonym keynote presented in the *International Symposium on Molecular Logic and Computational Synthetic Biology 2018*, traces an informal roadmap on Dynamic Logic (DL) field, focusing on its versatility and resilience to be adjusted and adopted in a wide class of application domains and computational paradigms. The exposition argues the room for developments on tagging DL to the analysis of synthetic biologic domain.

1 Introduction

Dynamic Logic [8] was introduced in the 70's by Pratt in [27] as a suitable logic to reason about, and verify, classic imperative programs. Since then, it evolved to an entire family of logics, which became increasingly popular for assertional reasoning about a wide range of systems and scenarios.

This talk guides an overview on this path prepared to the broad audience of this symposium, with interests and backgrounds ranging from formal Logics, control and systems theory to Synthetic Biology. Rather than to introduce technical aspects on the mentioned formalism, this presentation aims to raise the attention of the reader to the 'camaleonic' nature of DL, on its adoption on the verification of novel computational domains and paradigms, and in the way it can be yet extended to fit on the new challenges of synthetic biology.

This exposition starts by revisiting the roots of the topic, namely by (i) the generic ideas of the calculus of Floyd and Hoare on classic imperative programs and, by (ii) introduce Modal Logic, with its Kripke semantics, as the natural formalism to reason about state transition systems. Then, recalling the seminal ideas of Pratt of using a modal logic to perform Floyd-Hoare reasoning, we briefly introduce the propositional and first order versions of DL (see [28] for an historical perspective on the development of the topic).

Then, we overview some of contributions on the topic developed by our group. The dynamisation method [21,22] contributed on this direction with a systematic procedure to construct Multi-valued Dynamic Logics able to handle systems where the uncertainty is a prime concern. The method is parametric, and follows our own pragmatic approach to the application of logics to a wide range of complex computational systems: on the place of defining a dedicated logic for each

© Springer Nature Switzerland AG 2019
M. Chaves and M. A. Martins (Eds.): MLCSB 2018, LNCS 11415, pp. 18–33, 2019.
https://doi.org/10.1007/978-3-030-19432-1_2

specific application, we develop parametric methods to derive logics tailored to each situation or domain. The specificities involved in each situation should be taken in account on the definition of the parameter adopted for each derivation. The method reviewed in this talk, generates logics suitable to deal with systems involving graded computations. The grading of these logics is reflected in the costs, weight and certainty degrees of programs; but also in the assertions we can do, due to their multi-valued semantics (rather than the standard bivalent one). Beyond of standard Propositional Dynamic Logic [8], we can capture with this method, for instance, the Fuzzy Dynamic Logics presented in [12,16]. But other logics capable to reason with systems involving resource consuming computations, or assertions graded in discrete truth spaces, can also be achieved as well (cf. [21]).

This generic method have been also adjusted to build multi-valued variants of other families of logics. We discuss in this talk two possible specialisations: one to reason with systems involving knowledge - with its tuning to a method to build Multi-valued Dynamic Epistemic Logics (developed in collaboration with Martins and Benevides [2]; and, another one. to reason on weighted programs on means of intervals of weight, rather than points (developed in collaboration with Santiago, Martins and Bedregal [29,30]).

2   The Seminal Roots

Floyd-Hoare Calculus

As mentioned above, the works of Floyd and Hoare were determinant on the adventure of the formal verification discipline in software engineering. The standard concept of software corrections emerged from the ideas of [7,11], by means of the notion of Hoare triple:

$$\{\phi\}\,\pi\,\{\varphi\}$$

Formally, a triple $\{\phi\}\,\pi\,\{\varphi\}$ is valid if any terminating execution of π from a state satisfying ϕ, results in a state satisfying φ. Actually this notion of the program correctness w.r.t. a specification underlies, not only the modern techniques of software verification, but also the principles design-by-contract development and specification methods based in the state transitions with pre and post conditions. The Floyd-Hoare logic (Fig. 2) is a syntactic calculus to prove the correctness of a complex program by decomposing it into simpler ones. The intuitions for the set of axioms and inference rules is easy. For instance, the axiom (**assign**) just states that a condition φ is satisfied after an assignment $x := e$, whenever before of this assignment, the formula obtained by replacing in φ all the occurrences of x by the expression e, was already true. Axiom (**empty**) is also natural, since the program *skip* does not change states. As in the other natural deduction systems, the idea of this calculus is to decompose the proof of compound programs, into a set of simpler proof obligations, by creating a proof tree which leafs are axioms. This is clearly reflected in the (**comp**) and (**if_then_else**) inference rules. The rule (**weak**) allows to manipulate the triple conditions by strengthening preconditions and weakening postconditions. Using this rules we are able to validate Hoare triples. For instance,

$$\{x = 1\}\text{if } x < 2 \text{ then } x := x + 1 \text{ else } x := x * x\{x = 2\}$$

can be proved with the deduction:

$$\cfrac{\cfrac{\overline{\{x=1\}x:=x+1\{x=2\}}}{\{x=1\wedge x<2\}x:=x+1\{x=2\}} \qquad \overline{\{x=1\wedge x\geq 2\}x:=x*x\{x=2\}}}{\{x = 1\}\text{if } x < 2 \text{ then } x := x + 1 \text{ else } x := x * x\{x = 2\}}$$

The left leaf is closed by axiom (**assign**), since $(x = 2)[x + 1/x] \Leftrightarrow x = 1$. For the right one, we just have to note that $x = 1 \wedge x < 2 \Leftrightarrow false$, and therefore the triple is vacuously satisfied, since there is no any state satisfying the precondition $false$.

Axioms:

$$(\textbf{assign})\cfrac{}{\{\varphi[e/x]\}\,x := e\,\{\varphi\}} \qquad\qquad (\textbf{empty})\cfrac{}{\{\phi\}\,skip\,\{\phi\}}$$

Inference rules:

$$(\textbf{weak})\cfrac{\phi \to \phi' \quad \{\phi'\}S\{\varphi'\} \quad \varphi' \to \varphi}{\{\phi\}S\{\varphi\}} \qquad (\textbf{comp})\cfrac{\{\phi\}S\{\xi\} \quad \{\xi\}T\{\varphi\}}{\{\phi\}S;T\{\varphi\}}$$

$$(\textbf{if\_then\_else})\cfrac{\{\phi \wedge \alpha\}S_1\{\varphi\} \quad \{\phi \wedge \neg\alpha\}S_2\{\varphi\}}{\{\phi\}\text{if } \alpha \text{ then } S_1 \text{ else } S_2\{\varphi\}}$$

Fig. 1. Fragment of the Floyd-Hoare Calculus

Modal Logic

The long tradition in the study of logics to reasoning in scenarios involving change, come since the age of Aristotle. This family of logics, known as Modal logics represents a classic topic in Logic and Philosophy. The developments of Kripke in the 60's in semantics for these logics, based in transition structures, endow such formalisms with the suitable ingredients to reasoning about state-based systems. This section briefly review the basic definition of propositional multi-modal logic.

Signature for of this logic are pairs (Prop, A) where Prop, A are disjoint sets of propositions, and modalities. The (Prop, A)-formulas are defined by the grammar

$$\varphi ::= p \mid \langle a\rangle\varphi \mid [a]\varphi \mid \neg\varphi \mid \varphi \vee \varphi \mid \varphi \wedge \varphi$$

where $p \in$ Prop and $a \in A$.

Models of this logic are state transition structures, with propositions locally assigned to states. Formally, a (Prop, A)-model is a tuple $M = (W, V, R)$ where

- W is a set
- $V : \text{Prop} \to \mathcal{P}(W)$ is a function
- $R = (R_a \subseteq W \times W)_{a \in A}$ is an A-family of binary relations

Finally, we recall the notion of modal satisfaction. The satisfaction of a (Prop, A)-formula φ in a state w of a (Prop, A)-model M is recursively defined as follows:

- $M, w \models p$ iff $w \in V(p)$
- $M, w \models \langle a \rangle \varphi$ iff there is a $w' \in W$ such that $(w, w') \in R_a$ and $M, w' \models \varphi$
- $M, w \models [a]\varphi$ iff for any $w' \in W$ such that $(w, w') \in R_a$ we have $M, w' \models \varphi$
- $M, w \models \neg\varphi$ iff it is false that $M, w \models \varphi$
- $M, w \models \varphi \wedge \varphi'$ iff $M, w \models \varphi$ and $M, w \models \varphi'$
- $M, w \models \varphi \vee \varphi'$ iff $M, w \models \varphi$ or $M, w \models \varphi'$

Propositional Dynamic Logic

Being programs a paradigmatic example of state-transition systems, modal logic emerged as natural formalism to reason about it. Particularly, it provided solid theoretic field, to support the verifications in Floyd-Hoare triples. Moreover, as observed by V. Pratt in the seminal work [27] Floyd-Hoare logic is purely syntactic, and Modal logic can be considered as an alternative to Floyd-Hoare logic.

In a first view, the multi-modal logic presented above would be enough to reason about programs, by considering the class of possible programs as the set of modalities. Fortunately programs are structured terms. This allows us to deal with these objects in a systematic way, a key factor on the definition of dynamic logics. Assuming a set of atomic programs Π, the universe of the (composed) programs can be defined with the following grammar:

$$\pi ::= \pi_0 \mid \pi + \pi \mid \pi; \pi \mid \pi^* \mid ?\chi$$

for $\pi_0 \in \Pi$ and χ a formula in the logic. The connectives of the terms are the usual Kleene operators, namely $+$ represents the non-deterministic choice, ; the sequential composition and $*$ the reflexive iterative operator. Additionally we have the operator ? for tests, that is necessary to represent conditionals. Note that this grammar actually provides an abstract computational language, able to represent the standard imperative language commands. For instance we have that **if** χ **then** π **fi** $\equiv (?\chi; \pi) + (?\neg\chi)$, that **if** χ **then** π **else** π' **fi** $\equiv (?\chi; \pi) + (?\neg\chi; \pi')$ and that **while** χ **do** π **od** $\equiv (?\chi; \pi)^*; ?\neg\chi$.

Fixing this abstract model of computation, we are in condition to adjust multi-modal logic into a formalism to reasoning about programs. Firstly, signatures are pairs (Prop, Π) where Prop is a set of propositions and Π is a set of atomic programs names. Models are Kripke structures tuples (W, V, R) where:

- W is a set
- $V : \text{Prop} \to \mathcal{P}(W)$ is a function
- $R = (R_\pi \subseteq W \times W), \pi \in \Pi$

Observe that these models only interprets atomic programs, since $\overline{R}$ can be extended to the interpretation of composed programs, with the usual relational operators. Namely, we have $\overline{R}(\pi_0) = R_{\pi_0}$, $\overline{R}(\pi + \pi') = \overline{R}(\pi) \cup \overline{R}(\pi')$, $\overline{R}(\pi; \pi') = \overline{R}(\pi) \cdot \overline{R}(\pi')$ and $\overline{R}(\pi^*) = \overline{R}(\pi)^* = \bigcup_{n \in \mathbb{N}} \overline{R}(\pi^n)$, where $\pi^{n+1} = \pi; \pi^n$. Finally we have the interpretations of tests as the co-reflexive $\overline{R}(\chi?) = \{(w, w) | M, w \models \chi\}$. Now, we defined the satisfaction relation as above, just replacing the cases of modal operators by

- $M, w \models \langle \pi \rangle \varphi$ iff there is a $w' \in W$ such that $(w, w') \in \overline{R}_\pi$ and $M, w' \models \varphi$;
- $M, w \models [\pi] \varphi$ iff for any $w' \in W$ such that $(w, w') \in \overline{R}_\pi$ we have $M, w' \models \varphi$.

Shifting to the First-Order Case

As suggested, propositional dynamic logic provides the essential machinery to reason about abstract programs. The regular modalities reflect the abstract structure of the programs control, where the standard imperative commands can be easily accommodated. We observe here that above freedom on what an atomic program is a key factor to the versatility of this logic, to be adapted to new computational domains. Let us firstly focus in the verification of a classic imperative programs. For this case atomic programs are, naturally, variables assignments. As usual, the states in our models should correspond to valuations of program data variables. Hence, the atomic propositions used in the propositional case are here replaced by data predicates. For sake of simplicity, we assume that all the programs variables are numerical $\mathbb{R}$ variables.

Formally, signatures are sets of data variables Var. The set of programs is defined as in the propositional case, but considering assignments $x := \theta$, with θ a term defined with Var and the arithmetic operations $\{+, -, \times, \cdots\}$, on place of atomic programs $\pi \in \Pi$. As mentioned, semantic states are variables assignments $w \in \mathbb{R}^{\text{Var}}$. The interpretation of programs is now given by an interpretation $\rho \subseteq \mathbb{R}^{\text{Var}} \times \mathbb{R}^{\text{Var}}$ exactly defined as the propositional $\bar{R}$, but considering the interpretation of base programs $\rho_{x := \theta} = \{(u, v) | v(x) = \theta$ and for any $y \in \mathcal{V} \setminus \{x\}, u(y) = v(y)\}$.

Hence, we can use this modal logic to support the verification of Floyd-Hoare triples. For instance the validity of formula

$$x = 1 \rightarrow [(x < 2)?; x := x + 1 + (\neg (x < 2))?; x := x * x]x = 2$$

or, equivalently

$$x = 1 \rightarrow [\text{if } x < 2 \text{ then } x := x + 1 \text{ else } x := x * x]x = 2$$

corresponds to the verification of the triple

$$\{x = 1\}\text{if } x < 2 \text{ then } x := x + 1 \text{ else } x := x * x\{x = 2\}$$

done above. This is an useful fact that relates Floyd-Hoare logic and first-order dynamic logic: for any Floyd Hoare triple $\{\psi\}\pi\{\varphi\}$, $\{\psi\}\pi\{\varphi\}$ is verified iff the formula $\psi \rightarrow [\pi]\varphi$ is valid.

Note that this principle can be extended to other variants of Hoare and dynamic logics. Whenever a new dynamic logic is defined, a new Floyd-Hoare is created for free.

Less Conventional Variants

As stated in the introduction, the resilience of dynamic logic on being adjusted to new computational paradigms and domains is a key factor for its adoption in a wide multitude of contexts. Actually, the way we construct the first-order dynamic logic from its propositional version, by preserving all of its structure, with the exception of its atomic programs (and respective interpretation), not only justify the big family of dynamic logics we have today, but opens the door for further versions and variants. Actually, as it will be discussed, the DL adequacy and resilience on being adapted to a wide range of computational systems, relies on real understanding of what is the nature of the atomic programs involved in each context. On place of considering programs as the standard variables assignments, we can consider, for instance systems of differential equations flowing in a given domain (e.g. a time constraint or a data predicate). This is the base idea of differential dynamic logic of Platzer [26]. By considering as atomic programs these evolutions, we are in the presence of a logic to reason and verify continuous systems. But if we consider also discrete assignments we have a suitable logic to reason in hybrid systems.

A logic to reason about quantum programs and quantum protocols can be also achieved if we consider, as basic programs, quantum measurements and unitary transformations. Such is the idea behind the works of Smets and Baltag in Quantum Dynamic Logic [1]. The game logics of Parikh [25] and the Dynamic Epistemic logics (revisited bellow) [6] are two well established logic fields, where the same analogy can be done.

3 Parametric Generation of Dynamic Logics

This section overviews the dynamisation method, a systematic method to construct Multi-valued Dynamic Logics that we introduced in [21,22]. This method is parametrized by an action lattice [13]. Despite of its distinct original purposes, this algebraic structure showed to be very useful in the context of our work, on providing a generic support for the computational space (as a Kleene algebra) and for the truth spaces (as residuated lattice) of the logics build trough our constructions.

Definition 1 ([13]). *An* action lattice *is a tuple*

$$\mathbf{A} = (A, +, ;, 0, 1, *, \rightarrow, \cdot)$$

where A is a set, 0 and 1 are constants, $*$ is an unary operation in A and $+, ;, \rightarrow$ and $\cdot$ are binary operations in A satisfying the axioms enumerated in Fig. 1, where the relation $\leq$ is induced by $+$: $a \leq b$ iff $a + b = b$.

As discussed bellow, the structure of an action lattice plays a double role in our method: it will support the model for computations, and of truth space. The operation $+$, plays a double role, the non-deterministic choice, in the interpretation of programs, and the logical disjunction, in the interpretation of sentences.

$$a + (b + c) = (a + b) + c \qquad (1)$$
$$a + b = b + a \qquad (2)$$
$$a + a = a \qquad (3)$$
$$a + 0 = 0 + a = a \qquad (4)$$
$$a;(b;c) = (a;b);c \qquad (5)$$
$$a; 1 = 1; a = a \qquad (6)$$
$$a;(b + c) = (a;b) + (a;c) \qquad (7)$$
$$(a + b); c = (a;c) + (b;c) \qquad (8)$$
$$a; 0 = 0; a = 0 \qquad (9)$$
$$1 + a + (a^*; a^*) \leq a^* \qquad (10)$$

$$a; x \leq x \Rightarrow a^*; x \leq x \qquad (11)$$
$$x; a \leq x \Rightarrow x; a^* \leq x \qquad (12)$$
$$a; x \leq b \Leftrightarrow x \leq a \to b \qquad (13)$$
$$a \to b \leq a \to (b + c) \qquad (14)$$
$$(x \to x)^* = x \to x \qquad (15)$$
$$a \cdot (b \cdot c) = (a \cdot b) \cdot c \qquad (16)$$
$$a \cdot b = b \cdot a \qquad (17)$$
$$a \cdot a = a \qquad (18)$$
$$a + (a \cdot b) = a \qquad (19)$$
$$a \cdot (a + b) = a \qquad (20)$$

Fig. 2. Axiomatisation of action lattices

Operations $*$ and $;$ are taken to interpret the iterative application and sequential composition of actions and, the operations $\to$ and $\cdot$ interpret the logical implication and conjunction.

We explore [21] an extensive set of action lattice. Here we will just recall four of them. Firstly, we consider the two elements boolean algebra

$$\mathbf{2} = (\{\top, \bot\}, \vee, \wedge, \bot, \top, *, \to, \wedge)$$

with the standard boolean connectives and with $\top^* = \bot^* = \top$. Moreover, by explicitly introducing a denotation for a truth value *unknown*, we can consider the three elements linear lattice

$$\mathbf{3} = (\{\top, u, \bot\}, \vee, \wedge, \bot, \top, *, \to, \wedge)$$

where

$\vee$	$\bot$	u	$\top$		$\wedge$	$\bot$	u	$\top$		$\to$	$\bot$	u	$\top$		$*$	
$\bot$	$\bot$	u	$\top$		$\bot$	$\bot$	$\bot$	$\bot$		$\bot$	$\top$	$\top$	$\top$		$\bot$	$\top$
u	u	u	$\top$		u	$\bot$	u	u		u	$\bot$	$\top$	$\top$		u	$\top$
$\top$	$\top$	$\top$	$\top$		$\top$	$\bot$	u	$\top$		$\top$	$\bot$	u	$\top$		$\top$	$\top$

In order to consider a linear discrete lattice with a finite number of points, we can consider *Wajsberg hoops* [3] enriched with a suitable star operation. For a fix natural $k > 0$ and a generator a, we define the structure $\mathbf{W}_k = (W_k, +, ; , 0, 1, ^*, \to, \cdot)$, where $W_k = \{a^0, a^1, \cdots, a^k\}$, $1 = a^0$ and $0 = a^k$, and for any $m, n \leq k$, $a^m + a^n = a^{min\{m,n\}}$, $a^m; a^n = a^{min\{m+n,k\}}$, $(a^m)^* = a^0$, $a^m \to a^n = a^{max\{n-m,0\}}$ and $a^m \cdot a^n = a^{max\{m,n\}}$. For instace, the underlying order of the Wajsberg hoop $\mathbf{W}_5$ is $\mathbf{W}_5$ is $a^5 < a^4 < a^3 < a^2 < a^1 < a^0$.

Moreover, we can also consider continuous structures for the truth degrees and weight for our logics. For instance, the Łukasiewicz arithmetic lattice is the structure

$$\mathbf{L} = ([0, 1], max, \odot, 0, 1, *, \to, min)$$

where $x \to y = min(1, 1 - x + y)$, $x \odot y = max(0, y + x - 1)$ and $x^* = 1$.

Now, fixing an action lattice $\mathbf{A} = (A, +, ;, 0, 1, *, \rightarrow, \cdot)$ as parameter, we will construct the multi-valued dynamic logic $\mathcal{DL}(\mathbf{A})$ (as proposed in [17]). Signatures of $\mathcal{DL}(\mathbf{A})$ are pairs (Π, Prop) where Π denotes the set of atomic computations and Prop the set of propositions. Then, the *set of Π-programs* $\text{Prg}(\Pi)$, are defined by the grammar

$$\pi ::= \pi_0 \mid \pi; \pi \mid \pi + \pi \mid \pi^*, \text{ where } \pi_0 \in \Pi$$

Given a signature (Π, Prop), the set of formulas $\text{Fm}^{\mathcal{DL}}(\Pi, \text{Prop})$ is given by the grammar

$$\rho ::= \top \mid \bot \mid p \mid \rho \vee \rho \mid \rho \wedge \rho \mid \rho \rightarrow \rho \mid \rho \leftrightarrow \rho \mid \langle \pi \rangle \rho \mid [\pi] \rho$$

with $p \in \text{Prop}$ and $\pi \in \text{Prg}(\Pi)$.

Now we have to introduce the models for $\mathcal{DL}(\mathbf{A})$. As expected, graded computations will be interpreted in state transition systems with weights in the transitions, usually represented by adjency matrices. On this view, our method takes advantage of the Conway matricial constructions over Kleene algebras i.e. in the structure

$$\mathbb{M}_n(\mathbf{A}) = (M_n(\mathbf{A}), +, ;, 0, 1, *)$$

defined as in [4,14]. Namely with:

- $M_n(\mathbf{A})$ is the space of $(n \times n)$-matrices over $\mathbf{A}$.
- for any $A, B \in M_n(\mathbf{A})$, define $M = A + B$ by $M_{i,j} = A_{i,j} + B_{i,j}$, $i, j \leq n$.
- for any $A, B \in M_n(\mathbf{A})$, define $M = A ; B$ by $M_{i,j} = \sum_{k=1}^{n}(A_{i,k}; B_{k,j})$ for any $i, j \leq n$.
- **1** and **0** are the $(n \times n)$-matrices defined by $\mathbf{1}_{i,j} = \begin{cases} 1 & \text{if } i = j \\ 0 & \text{otherwise} \end{cases}$ and $\mathbf{0}_{i,j} = 0$, for any $i, j \leq n$.
- for any $M = [a] \in \mathbb{M}_1(\mathbf{A})$, $M^* = [a^*]$;
  for any $M = \left[\begin{array}{c|c} A & B \\ \hline C & D \end{array}\right] \in M_n(\mathbf{A})$, $n > 1$, where A and D are square matrices, define

$$M^* = \left[\begin{array}{c|c} F^* & F^*; B; D^* \\ \hline D^*; C; F^* & D^* + (D^*; C; F^*; B; D^*) \end{array}\right]$$

where $F = A + B; D^*; C$. Note that this construction is recursively defined from the base case (where $n = 2$) where the operations of the base action lattice $\mathbf{A}$ are used.

As showed in [14], the structure $\mathbb{M}_n(\mathbf{A})$ is also a Kleene algebra, and therefore, figures as a suitable space to represent, manipulate and interpret programs. Enriching the interpretation of basic programs with graded interpretations for the propositions, we get the models for a signature (Π, Prop). Formally, the $\mathcal{DL}(\mathbf{A})$ models for (Π, Prop) are tuples

$$\mathcal{A} = (W, V, (\mathcal{A}_\pi)_{\pi \in \Pi})$$

where W is a finite set (of states), $V : \text{Prop} \times W \to A$ is a function, and $\mathcal{A}_\pi \in \mathbb{M}_n(\mathbf{A})$, with n standing for the cardinality of W.

As expected, the interpretation of a program $\pi \in \text{Prg}(\Pi)$ in a model $\mathcal{A} \in \text{Mod}^{\mathcal{DL}}(\Pi, \text{Prop})$ is recursively defined, from the set of atomic programs $(\mathcal{A}_\pi)_{\pi \in \Pi}$, with $\mathcal{A}_{\pi;\pi'} = \mathcal{A}_\pi ; \mathcal{A}_{\pi'}, \mathcal{A}_{\pi+\pi'} = \mathcal{A}_\pi + \mathcal{A}_{\pi'}$ and $\mathcal{A}_{\pi*} = \mathcal{A}_\pi^*$ together with the constants interpretations $\mathcal{A}_1 = \mathbf{1}$ and $\mathcal{A}_0 = \mathbf{0}$.

The reader can easily observe that the models of $\mathcal{DL}(\mathbf{2})$ corresponds exactly to the standard PDL. More interesting instantiations can be found in [21].

In order to illustrate the running concepts, let us consider the consider the $(\{p, q\}, \{\pi, \pi'\})$-model $\mathcal{A} = (\{s_1, s_2\}, V, (\mathcal{A}_p)_{p \in \{\pi, \pi'\}})$ of $\mathcal{DL}(L)$ with $V(p, s_1) = 0.1$, $V(q, s_1) = 0.5$, $V(p, s_2) = \frac{\pi}{4}$ and $V(q, s_2) = 0.75$ and

$$
\mathcal{A}_\pi : \quad \underset{s_1}{\bigcirc} \xrightarrow{\frac{\sqrt{2}}{3}} \overset{0.7}{\underset{s_2}{\circlearrowright}} \begin{bmatrix} 0 & \frac{\sqrt{2}}{3} \\ 0 & 0.7 \end{bmatrix} \qquad \mathcal{A}_{\pi'} : \quad \underset{s_1}{\bigcirc} \overset{\frac{\sqrt{2}}{2}}{\underset{\frac{\sqrt{3}}{2}}{\rightleftarrows}} \overset{0.5}{\underset{s_2}{\circlearrowright}} \begin{bmatrix} 0 & \frac{\sqrt{2}}{2} \\ \frac{\sqrt{3}}{2} & 0.5 \end{bmatrix} \tag{21}
$$

Then, for instance the program $\mathcal{A}_{\pi+\pi'}$ is interpreted by

$$
\mathcal{A}_{\pi+\pi'} = \max(\mathcal{A}_\pi, \mathcal{A}_{\pi'}) = \max\left(\begin{bmatrix} 0 & \frac{\sqrt{2}}{3} \\ 0 & 0.7 \end{bmatrix}, \begin{bmatrix} 0 & \frac{\sqrt{2}}{2} \\ \frac{\sqrt{3}}{2} & 0.5 \end{bmatrix} \right) = \begin{bmatrix} 0 & \frac{\sqrt{2}}{2} \\ \frac{\sqrt{3}}{2} & 0.7 \end{bmatrix} \tag{22}
$$

The last ingredient for the definition of $\mathcal{DL}(L)$ is the graded satisfaction. Here, on place of being a satisfaction relating each state with the formulas there satisfied, we have a function that assigns the 'satisfaction degree' of a formula in a given state. The operations of the action lattices have to play the truth space role, on the interpretation of logic connectives. Formally, the graded satisfaction relation for a model $\mathcal{A} \in \text{Mod}^{\mathcal{DL}}(\Pi, \text{Prop})$, with $\mathbf{A}$ complete, consists of a function

$$
\models : W \times \text{Fm}^{\mathcal{DL}}(\Pi, \text{Prop}) \to A
$$

recursively defined as follows:

- $(w \models \top) = \top$
- $(w \models \bot) = \bot$
- $(w \models p) = V(p, w)$, for any $p \in \text{Prop}$
- $(w \models \rho \wedge \rho') = (w \models \rho) \cdot (w \models \rho')$
- $(w \models \rho \vee \rho') = (w \models \rho) + (w \models \rho')$
- $(w \models \rho \to \rho') = (w \models \rho) \to (w \models \rho')$
- $(w \models \rho \leftrightarrow \rho') = (w \models \rho \to \rho'); (w \models \rho' \to \rho)$
- $(w \models \langle \pi \rangle \rho) = \sum_{w' \in W} (\mathcal{A}_\pi(w, w'); (w' \models \rho))$
- $(w \models [\pi] \rho) = \prod_{w' \in W} (\mathcal{A}_\pi(w, w') \to (w' \models \rho))$

We say that ρ is *valid* when, for any model $\mathcal{A}$, and for each state $w \in W$, $(w \models \rho) = \top$.

Returning to our running example, we can calculate the satisfaction degree of the formula $\langle \pi + \pi' \rangle (p \to q))$ in the state s_1 as follows:

$$(s_1 \models \langle \pi + \pi' \rangle (p \to q)) = \max(0 \odot (0.1 \to 0.5), \frac{\sqrt{2}}{2} \odot (0.75 \to \frac{\pi}{4}))$$

$$= \frac{\sqrt{2}}{2} \odot (0.75 \to \frac{\pi}{4})$$

$$= \frac{\sqrt{2}}{2} \odot \min(1, 1 - 0.75 + \frac{\pi}{4})$$

$$= \frac{\sqrt{2}}{2}$$

Therefore, we conclude with a degree of certainty $\frac{\sqrt{2}}{2}$ that, after executing $\pi + \pi'$ from the state s_1, we have $p \to q$.

Reasoning with Systems Involving Knowledge

The complexity of the current information systems, involving processes with complex network of heterogeneous learning agents, raises for further generalisations of Multi-agent Epistemic Logics, including weighted versions. Hence, the building logics on-demand principle, inherent to dynamisation, appear as an adequate technique to be used is this domain. In this section, we review a variant of dynamisation tailored to the generation of graded dynamic epistemic logics introduced in [2].

Firstly let us recall the basis of Multi-agents Epistemic Logic (DEL). Signature of DEL are pairs (Prop, Ag) where Prop is a set of propositions and Ag a finite set of agents. Note that this can be seen as propositional dynamic logic signatures which atomic programs are the agent knowledge relations. The (Prop, Ag) formulas of DEL are defined by the grammar

$$\varphi ::= p \mid \top \mid \neg\varphi \mid \varphi_1 \wedge \varphi_2 \mid \varphi_1 \vee \varphi_2 \mid K_a\varphi \mid B_a\varphi \mid C_G\varphi$$

where $p \in$ Prop, $a \in$ Ag and $G \subseteq$ Ag. The intuitive meaning of the epistemic modalities is the following: $K_a\varphi$ means that agent a knows φ; $B_a\varphi$ means that agent a believes that φ; and the common knowledge operator $C_G\varphi$ - means that all the members of the group of agents G knows φ and each member of the group knows that all the members of the groups know φ, etc.

The models are just special models of PDL. Formally, *multi-agent epistemic model* is a tuple $\mathcal{E} = (W, (R_a)_{a\in Ag}, V)$ defined as in PDL but assuming that, for any agent $a \in$ Ag, R_a is an equivalence relation. The interpretation of knowledge modalities is defined by

- $\mathcal{M}, s \models K_a\phi$ iff for all $s' \in S : sR_a s' \Rightarrow \mathcal{M}, s' \models \phi$
- $\mathcal{M}, s \models B_a\phi$ iff there is an $s' \in S$ such that $sR_a s'$ and $\mathcal{M}, s' \models \phi$
- $\mathcal{M}, s \models C_G\phi$ iff for all $s' \in S, sR_G^* s' \Rightarrow \mathcal{M}, s' \models \phi$

The similarities with PDL are straightforward. Modality K_a corresponds to the modality $[a]$ for an atomic program a. Its dual, the modality B_a, corresponds to the modality $\langle a \rangle$ for an atomic program a. Modality C_G is captured by the modality $[(\sum_{a \in G} a)^*]$.

In order to get some intuitions on this logic, let us recall the well know example of the envelops used in [6]. Three envelopes containing **0**, **1** and **2** euros are given to the agents **ana**, **bob** and **clara**. Each agent just knowns the content of her envelop. Using proposition Prop $= \{E_x | E \in \{1,2,3\}, x \in \{a,b,c\}\}$ referring that "agent x has envelop E", and representing states by the order of envelops, e.g. the state 012 represents the case that agent **a** has **0**, agent **b** has **1** and **c** has **2**, we can represent epistemic state of each agent as follows[1] (Fig. 3):

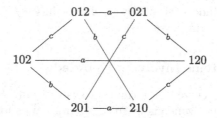

Fig. 3. anna's, bob's and clara's epistemic model [6]

Hence, we have, for instance, that $012 \models B_b 0_a$ and $012 \models B_a K_c 2_c$ hold. Redefining our dynamisation method for this specific seetings we obtain a method to build graded dynamic epistemic logics. The satisfaction relation for the epistemic modalities takes now the form:

- $(w \models K_a \varphi) = \bigwedge_{w' \in W} \left(R_a(w,w') \to (w' \models \varphi) \right)$
- $(w \models B_a \varphi) = \bigvee_{w' \in W} \left(R_a(w,w'); (w' \models \varphi) \right)$
- $(w \models C_G \varphi) = \bigwedge_{w' \in W} \left(R_G^*(w,w') \to (w' \models \varphi) \right)$

These logics are prepared to deal with agents with graded beliefs (on place of bivalent ones). Let us revisit the example above, by supposing that the agent **ana** 'suspect' that the envelop of **bob** has a higher amount than the one of herself. In a scale from 0 to 5, her belief is 4; Conversely, her belief that the envelop received by **bob** has a smaller value is 1. The epistemic perception of **ana** is depicted in the following picture. Again, we omit the reflexive loops in the picture (with value 5) (Fig. 4):[2]

Reasoning with Interval Approximations

There are some situations where only approximations for the transition weights are possible (e.g. when dealing weights over irrational numbers, we have no

---

[1] We omit the reflexive loops in the picture.
[2] The complete treatment of this illustration is in [2];.

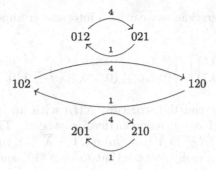

Fig. 4. anna's beliefs

machine representation of transition weights; or due impreciseness in some measurements). On this purpose, for the specific case of Fuzzy Dynamic logic, we adjusted the dynamisation constructions to deal with intervals, rather than points. This results in a new family of dynamic logics whose assertions, and the satisfaction outcomes, are also intervals. This section informally overviews our work in Interval Dynamic Logic presented in [29, 30]. The presentation is guided to the case L but the same principle can be extended to other continuous action algebras.

In the sequel, for any closed interval X, we use $\underline{X}$ and $\underline{Y}$ to denote its left and right bounds, i.e. for $X = [a, b]$, $\underline{X} = a$ and $\underline{Y} = b$.

Our first concern is about the structure to interpret such kind of programs. The following result presented in [29] provides a Kleene algebra for that end:

Theorem 1 ([29]).

$$K(\widehat{L}) = (\mathbb{U}, Max, \bigodot, [0, 0], [1, 1], \widehat{*})$$

where

- $\mathbb{U} = \{[a, b] \mid 0 \leq a \leq b \leq 1\}$
- $Max(X, Y) = [\max(\underline{X}, \underline{Y}), \max(\overline{X}, \overline{Y})]$
- $Min(X, Y) = [\min(\underline{X}, \underline{Y}), \min(\overline{X}, \overline{Y})]$
- $X \bigodot Y = [(\underline{X} \odot \underline{Y}), (\overline{X} \odot \overline{Y})] = [\max(0, \underline{X} + \underline{Y} - 1), \max(0, \overline{X} + \overline{Y} - 1)]$
- $X^{\widehat{*}} = [\underline{X}^*, \overline{X}^*] = [1, 1]$.

is a Kleene algebra.

For instance, we can consider interval approximations of the weight transition structure presented above as

$$\mathcal{A}_\pi : \quad \begin{array}{c} [0.7, 0.7] \\ \\ \boxed{s_1} \xrightarrow{[0.4, 0.5]} \boxed{s_2} \end{array} \quad \begin{bmatrix} (0, 0) & (0.4, 0.5) \\ (0, 0) & (0.7, 0.7) \end{bmatrix} \qquad \mathcal{A}_{\pi'} : \quad \begin{array}{c} [0.5, 0.5] \\ \\ \boxed{s_1} \underset{[0.7, 0.9]}{\overset{[0.6, 0.8]}{\rightleftarrows}} \boxed{s_2} \end{array} \quad \begin{bmatrix} (0, 0) & (0.6, 0.8) \\ (0.7, 0.9) & (0.5, 0.5) \end{bmatrix}$$

Using the $K(\widehat{L})$ operations, we can also interpret (composed) programs. For instance $\mathcal{A}_{\pi+\pi'}$ is

$$\max\left(\begin{bmatrix}(0,0) & (0.4,0.5)\\(0,0) & (0.7,0.7)\end{bmatrix}, \begin{bmatrix}(0,0) & (0.6,0.8)\\(0.7,0.9) & (0.5,0.7)\end{bmatrix}\right) = \begin{bmatrix}(0,0) & (0.6,0.8)\\(0.7,0.9) & (0.7,0.7)\end{bmatrix}$$

It is expected to extend the structure $K(\widehat{L})$ with an interpretation of the implication, in order to have an action lattice for intervals. The natural candidate is $X{\Rrightarrow}Y = [(\overline{X} \to \underline{Y}), (\underline{X} \to \overline{Y})] = [\min(1, 1 - \overline{X} + \underline{Y}), \min(1, 1 - \underline{X} + \overline{Y})]$. However, the reader can easily observe that axioms (13) and (15) does not hold in $\widehat{L}$ (c.f. [29] for a complete discussion). Hence, despite of its Kleene algebra structure, $\widehat{L}$ it is not an action lattice. We studied in [29], an weakness of action lattices, called *quasi-action lattice*, that capture $\widehat{L}$. Fortunately, this structures still have good properties to serve as parameter of dynamisation method. For instance, by considering a valuation $V : \text{Prop} \to \mathbb{U}$ with $V(p, s_1) = [0.1, 0.1]$, $V(q, s_1) = [0.5, 0.5]$, $V(p, s_2) = [0.7, 0.8]$ and $V(q, s_2) = [0.75, 0.75]$, we can calculate the degree of satisfaction of the sentence $\langle \pi + \pi' \rangle(p \to q)$ from the state s_1 as:

$(s_1 \models_{\widehat{L}} \langle \pi + \pi' \rangle(p \to q))$
$= \max([0,0] \odot ([0.1, 0.1] {\Rrightarrow} [0.5, 0.5]), [0.6, 0.8] \odot ([0.5, 0.5] {\Rrightarrow} [0.7, 0.8]))$
$= \max([0,0], [0.6, 0.8] \odot [0.5 \to 0.7, 0.5 \to 0.8])$
$= [0.6, 0.8] \odot [1, 1]$
$= [0.6, 0.8].$

4　Further Extensions and Applications?

As suggested along the paper, the pattern of *changing the atomic programs to adapt the computing paradigm* is not only recognised in our methods to build graded dynamic logics, but in most of variants of dynamic logics in the literature. This motivates our position that the shape of dynamic logic provides the *de facto* essence of what a logic for programs is. When invited to make a personal overview in dynamic logic in the International Symposium on Molecular Logic and Computational Synthetic Biology 2018, the authors main motivation was to open the discussion of what should be the suitable atomic programs, for a further dynamic logic tailored to synthetic biology. The same exercise have been done by the group on finding new dynamic logics for other domains and applications, including reactive processes [9,10,17], petri-nets with failures [15]. We have also explored this 'logic-on-demand' strategy in other modal logics. Our long term research in the parametric generation of hybrid logics [19,23,24] supports the formal development of a wide range of reconfigurable systems from the design to the verification stage [20]. Moreover, we extended the parametric generation of Dynamic Epistemic Logics in [18] by considering structured representation of states.

Exploiting the limits of our methods on building dynamic logics prepared to deal with paraconsistencies in behaviours or in knowledge acquirement, is a research line that we intend to develop. This will certainly be useful for the application domains of this symposium. The recent contributions within the group in paraconsistent hybrid logic [5] provides an interesting starting point for this agenda. Shifting the paraconsistency of atomic modalities to composed programs is, however, challenging. Specific questions as 'what is a paraconsistent program' should be answered. More precisely, the understanding of what is a paraconsistent execution of a program, if the paraconsistency is inherent to the atomic programs, or if it results from a 'paraconsistent control', due non conventional interpretation of the Kleene operators, are questions to be studied in this line.

Acknowledgements. The author would sincerely thanks the invitation of the *MLCSB'18* to present this personal perspective on Dynamic Logic, a topic with which he has been involved in the last years.

This work is financed by the ERDF – European Regional Development Fund through the Operational Programme for Competitiveness and Internationalisation - COMPETE 2020 Programme and by National Funds through the Portuguese funding agency, FCT - Fundação para a Ciência e a Tecnologia, within project POCI-01-0145-FEDER-016692 and UID/MAT/04106/2019, in a contract foreseen in nos. 4–6 of art. 23 of the DL 57/2016, changed by DL 57/2017.

References

1. Baltag, A., Smets, S.: Quantum logic as a dynamic logic. Synthese **179**(2), 285–306 (2011). https://doi.org/10.1007/s11229-010-9783-6
2. Benevides, M., Madeira, A., Martins, M.: A family of graded epistemic logics. Electr. Notes Theor. Comput. Sci. **338**, 45–59 (2018). https://doi.org/10.1016/j.entcs.2018.10.004
3. Blok, W.J., Ferreirim, I.M.A.: On the structure of hoops. Algebra Univers. **43**(2–3), 233–257 (2000). https://doi.org/10.1007/s000120050156
4. Conway, J.H.: Regular Algebra and Finite Machines. Printed in GB by William Clowes & Sons Ltd. (1971)
5. Costa, D., Martins, M.A.: Paraconsistency in hybrid logic. J. Log. Comput. **27**(6), 1825–1852 (2017). https://doi.org/10.1093/logcom/exw027
6. van Ditmarsch, H., van der Hoek, W., Kooi, B.: Dynamic Epistemic Logic. Synthese Library Series. Springer, The Netherland (2008). https://doi.org/10.1007/978-1-4020-5839-4
7. Floyd, R.W.: Assigning meanings to programs. In: Proceedings of Symposium on Applied Mathematics, vol. 19, pp. 19–32 (1967). http://laser.cs.umass.edu/courses/cs521-621.Spr06/papers/Floyd.pdf
8. Harel, D., Kozen, D., Tiuryn, J.: Dynamic Logic. MIT Press, Cambridge (2000)
9. Hennicker, R., Madeira, A.: Institutions for behavioural dynamic logic with binders. In: Hung, D.V., Kapur, D. (eds.) Theoretical Aspects of Computing - ICTAC 2017–14th International Colloquium, Hanoi, Vietnam, October 23–27, 2017, Proceedings. LNCS, vol. 10580, pp. 13–31. Springer, Cham (2017). https://doi.org/10.1007/978-3-319-67729-3_2
10. Hennicker, R., Madeira, A., Knapp, A.: A hybrid dynamic logic for event/data-based systems. In: Hähnle, R., van der Aalst, W. (eds.) FASE 2019. LNCS, vol. 11424, pp. 79–97. Springer, Cham (2019). https://doi.org/10.1007/978-3-030-16722-6_5

11. Hoare, C.A.R.: An axiomatic basis for computer programming. Commun. ACM **12**(10), 576–580 (1969). https://doi.org/10.1145/363235.363259
12. Hughes, J., Esterline, A.C., Kimiaghalam, B.: Means-end relations and a measureof efficacy. J. Logic Lang. Inf. **15**(1–2), 83–108 (2006). https://doi.org/10.1007/s10849-005-9008-4
13. Kozen, D.: On action algebras, manuscript. In: Logic and Flow of Information, Amsterdam (1991)
14. Kozen, D.: A completeness theorem for Kleene algebras and the algebra of regular events. Inf. Comput. **110**(2), 366–390 (1994)
15. Leandro Gomes, A.M., Benevides, M.: Logics for petri nets with propagating failures. FSEN19 - Fundamentals of Software Engineering. LNCS (in print)
16. Liau, C.: Many-valued dynamic logic for qualitative decision theory. In: Zhong, N., Skowron, A., Ohsuga, S. (eds.) New Directions in Rough Sets, Data Mining, and Granular-Soft Computing, 7th International Workshop, RSFDGrC 1999, Yamaguchi, Japan, November 9–11, 1999, Proceedings. LNCS, vol. 1711, pp. 294–303. Springer, Berlin (1999). https://doi.org/10.1007/978-3-540-48061-7-36
17. Madeira, A., Barbosa, L.S., Hennicker, R., Martins, M.A.: A logic for the stepwise development of reactive systems. Theor. Comput. Sci. **744**, 78–96 (2018). https://doi.org/10.1016/j.tcs.2018.03.004
18. Madeira, A., Benevides, M., Martins, M.: Epistemic logics with structured states. Electr. Notes Theor. Comput. Sci. (in print)
19. Madeira, A., Martins, M.A., Barbosa, L.S., Hennicker, R.: Refinement inhybridised institutions. Formal Asp. Comput. **27**(2), 375–395 (2015). https://doi.org/10.1007/s00165-014-0327-6
20. Madeira, A., Neves, R., Barbosa, L.S., Martins, M.A.: A method for rigorousdesign of reconfigurable systems. Sci. Comput. Program. **132**, 50–76 (2016). https://doi.org/10.1016/j.scico.2016.05.001
21. Madeira, A., Neves, R., Martins, M.A.: An exercise on the generation of many-valued dynamic logics. J. Log. Algebr. Meth. Program. **85**(5), 1011–1037 (2016). https://doi.org/10.1016/j.jlamp.2016.03.004. http://www.sciencedirect.com/science/article/pii/S2352220816300256
22. Madeira, A., Neves, R., Martins, M.A., Barbosa, L.S.: A dynamic logic for every season. In: Braga, C., Martí-Oliet, N. (eds.) Formal Methods: Foundations and Applications - 17th Brazilian Symposium, SBMF 2014, Maceió, AL, Brazil, September 29-October 1, 2014. Proceedings. LNCS, vol. 8941, pp. 130–145. Springer, Cham (2014). https://doi.org/10.1007/978-3-319-15075-8_9
23. Martins, M.A., Madeira, A., Diaconescu, R., Barbosa, L.S.: Hybridization of institutions. In: Corradini, A., Klin, B., Cîrstea, C. (eds.) Algebra and Coalgebra in Computer Science - 4th International Conference, CALCO 2011, Winchester, UK, August 30 - September 2, 2011. Proceedings. LNCS, vol. 6859, pp. 283–297. Springer, Heidelberg (2011). https://doi.org/10.1007/978-3-642-22944-2_20
24. Neves, R., Madeira, A., Martins, M.A., Barbosa, L.S.: Proof theory for hybrid(ised) logics. Sci. Comput. Program. **126**, 73–93 (2016). https://doi.org/10.1016/j.scico.2016.03.001
25. Parikh, R.: The logic of games and its applications. In: Selected Papers of the International Conference on "Foundations of Computation Theory" on Topics in the Theory of Computation, pp. 111–139. Elsevier North-Holland Inc, New York (1985). http://dl.acm.org/citation.cfm?id=4030.4037
26. Platzer, A.: Logical Foundations of Cyber-Physical Systems. Springer, Cham (2018). https://doi.org/10.1007/978-3-319-63588-0

27. Pratt, V.R.: Semantical considerations on floyd-hoare logic. In: 17th Annual Symposium on Foundations of Computer Science, Houston, Texas, USA, 25–27 October 1976, pp. 109–121. IEEE Computer Society (1976). https://doi.org/10.1109/SFCS.1976.27

28. Pratt, V.R.: Dynamic logic: A personal perspective. In: Madeira, A., Benevides, M.R.F. (eds.) Dynamic Logic. New Trends and Applications - First International Workshop, DALI 2017, Brasilia, Brazil, September 23–24, 2017, Proceedings. LNCS, vol. 10669, pp. 153–170. Springer, Cham (2017). https://doi.org/10.1007/978-3-319-73579-5_10

29. Santiago, R., Bedregal, B., Madeira, A., Martins, M.A.: On interval dynamic logic: introducing quasi-action lattices. Sci. Comput. Program. **175**, 1–16 (2019). https://doi.org/10.1016/j.scico.2019.01.007. http://www.sciencedirect.com/science/article/pii/S0167642319300103

30. Santiago, R.H.N., Bedregal, B., Madeira, A., Martins, M.A.: On interval dynamic logic. In: Ribeiro, L., Lecomte, T. (eds.) SBMF 2016. LNCS, vol. 10090, pp. 129–144. Springer, Cham (2016). https://doi.org/10.1007/978-3-319-49815-7_8

Oscillatory Behaviour
on a Non-autonomous Hybrid SIR-Model

Eugénio M. Rocha[(✉)][iD]

Center for Research and Development in Mathematics and Applications,
Department of Mathematics, University of Aveiro, 3810-193 Aveiro, Portugal
eugenio@ua.pt

Abstract. We study the impact of some abstract agent intervention on
the disease spread modelled by a SIR-model with linear growth infectivity. The intervention is meant to decrease the infectivity, which are
activated by a threshold on the number of infected individuals. The coupled model is represented as a nonlinear non-autonomous hybrid system. Stability and reduction results are obtained using the notions of
non-autonomous attractors, Bohl exponents, and dichotomy spectrum.
Numerical examples are given where the number of infected individuals
can oscillate around a equilibrium point or be a succession of bump functions, which are validated with a tool based on the notion of δ-complete
decision procedures for solving satisfiability modulo theories problems
over the real numbers and bounded δ-reachability. These findings seem
to show that hybrid SIR-models are more flexible than standard models
and generate a vast set of solution profiles. It also raises questions regarding the possibility of the agent intervention been somehow responsible
for the shape and intensity of future outbreaks.

Keywords: SIR-models · Hybrid systems · Stability

1 Introduction

Mathematically, the model of choice to represent the dynamics of the epidemic
is the SIR-model and its variants; introduced by Kermack and McKendrick [11].
Since then, the literature on the subject is quite vast. However, one of the key
issues in the subject is that the simplicity of (autonomous) SIR-models do not
produce solutions with complicate oscillatory behaviour.

Although less common, non-autonomous SIR-models have been introduced
and studied in the literature and may overcome partially such limitations, e.g.
see [2–4, 16, 18, 21, 23]. Usually these models introduce some type of seasonality
behaviour, for example, through a periodic infectivity function. Indeed, Bacaër
et al. [2] introduced a generalization of the basic reproduction number, and
Boatto et al. [4] considered a SIR-model with birth and death terms and time-
varying infectivity as a sinusoidal, showing that the (average) basic reproduction
number, the initial phase, the amplitude and the period are all relevant issues.

© Springer Nature Switzerland AG 2019
M. Chaves and M. A. Martins (Eds.): MLCSB 2018, LNCS 11415, pp. 34–55, 2019.
https://doi.org/10.1007/978-3-030-19432-1_3

Moreover, they show the existence of a periodic orbit. Bai et al. [3] studied a model with a seasonal contact rate and a staged treatment strategy, showing two different bistable behaviours under certain conditions: the stable disease-free state coexists with a stable endemic periodic solution, and three endemic periodic solutions coexist with two of them being stable. Kuniya [16] deals with an age-structured SIR epidemic model with time periodic coefficients, obtaining the basic reproduction number as the spectral radius of the next generation operator and showing that it plays the role of a threshold value for the existence of a nontrivial periodic solution; based on a Krasnoselskii fixed point theorem argument. Another approach to produce non-autonomous systems is to couple different SIR-models by a non-autonomous function, e.g. Rocha et al. [21] introduced a tuberculosis (TB) mathematical model, with 25 state-space variables where 15 are evolution disease states (EDSs), which takes into account the (seasonal) flux of populations between a high incidence TB country (A) and a host country (B) with low TB incidence, where (B) is divided into a community (G) with high percentage of people from (A) plus the rest of the population (C).

In this work, we consider an infectivity function which grows linearly (i.e. the most simple non-autonomous function), but we also study the effect of an agent intervention on the model in the form of some action policies. The policies are meant to reduce to decrease the infectivity, which are activated by a threshold on the number of infected individuals. Such approach turns the full model into a nonlinear non-autonomous hybrid system, see Sect. 2. Stability and reduction results are obtained using the notions of non-autonomous attractors, Bohl exponents, and dichotomy spectrum, which are presented in Sect. 3. In Sect. 4, we give some numerical examples, where the number of infected individuals can oscillate around a equilibrium point or be a succession of bump functions. The last example is quite interesting and raises questions regarding the possibility of the agent intervention time been somehow responsible for the shape and intensity of some future outbreaks. We end this work with some brief concluding remarks.

2   The Mathematical Model

2.1   The Class of Non-autonomous ODEs

Consider the basic SIR epidemic model together with a piecewise linear continuous infection coefficient β_ξ, described by

$$(a) \begin{cases} S' = \alpha R + (\zeta + \alpha)I - \beta_\xi IS, \\ I' = \beta_\xi IS - (\zeta + \alpha + \gamma)I, \\ R' = \gamma I - \alpha R, \end{cases} \quad \text{and} \quad (b)\ \beta'_\xi = \xi, \qquad (1)$$

where $\gamma > 0$, $\alpha \geq 0$, $\zeta \geq -\alpha$, and $\xi \in \mathbb{R}$ is a bifurcation parameter; e.g. for $\xi = 0$ the model is autonomous. The values $S(t), I(t), R(t)$ are, respectively, the number of healthy individuals (susceptible), infected individuals and recovered individuals; and α is a parameter of birth and death, γ is a recovery rate without possibility of re-infection, and ζ accounts for the rate of individuals that become healthy but may be re-infected in the future. We assume a (normalized) constant population $S + I + R = 1$, so the system (1)(a) evolves on the simplex defined by

$$\Sigma_1 = \{(S, I, R) \in \mathbb{R}^3 : S, I, R \geq 0, \ S + I + R = 1\},$$

meaning that system (1) may be written as

$$(a) \begin{cases} S' = \alpha(1 - S) + \zeta I - \beta_\xi IS, \\ I' = \beta_\xi IS - (\zeta + \alpha + \gamma)I, \\ (S(t), I(t), R(t)) \in \Sigma_1, \end{cases} \quad \text{and} \quad (b) \ \beta_\xi' = \xi, \quad (2)$$

For mathematical reasons, which will be clear in what follow, e.g. use of pullback limits and Bohl exponents, we work with an unbounded from below time interval $\overline{\mathcal{T}} = (-\infty, \mathcal{T}]$, for $\mathcal{T} > 0$, with an initial time $t_0 \in (0, \mathcal{T})$. For convenience, from now on, we use the notations $\overline{\mathcal{T}}_{t_0} = [t_0, \mathcal{T}]$, SIR($\xi$) to describe the set of Eq. (2) for a given parameter $\xi \in \mathbb{R}$, which account for the (linear) increase/decrease ratio of the disease.

Clearly, Eq. (2)(b) may be extended by using other growth functions, e.g. accounting for saturation phenomena, instead of the simple linear change in the infection coefficient. However, for the purpose of this work, such is enough in order to discover the main differences from the standard (autonomous) SIR-model, vastly used in the literature.

2.2 Non-autonomous Hybrid SIR-Models Generated by Simple Action Policies

In the model under study, we have two main entities, i.e. the natural evolution of the disease (nature) versus the evolution of the disease together with some abstract agent action with the purpose of reducing the transmission rate. Each will have a on/off-state, but makes sense to suppose nature is always in the on-state when the agent action is in the off-state, and vice-versa. Since they are complementary, we consider states in the viewpoint of the agent action. Further, to model the action from the agent, we assume that it depends on the current number of infected individuals $I(t)$ and has a maximum fixed time of intervention $T^* > 0$. For that, we establish two threshold values as triggers to the on/off-states, namely, $I_b \in (0, 1]$ and $I_s \in [0, 1)$. Then, the agent strategies considered are:

(S_0) the action starts at time $\tilde{t} \in \overline{\mathcal{T}}_{t_0}$, if it was in the off-state and $I(\tilde{t}) = I_b$, then stops at time $t = \tilde{t} + T^*$ (i.e. $I_b = 1$);

(S_1) the action starts at time $\tilde{t} \in \overline{\mathcal{T}}_{t_0}$, if it was in the off-state and $I(\tilde{t}) = I_b$, then stops at the first time $t > \tilde{t}$ with $I(t) = I_s$ (i.e. $T^* = +\infty$);

(S_2) the action starts at time $\tilde{t} \in \overline{\mathcal{T}}_{t_0}$, if it was in the off-state and $I(\tilde{t}) = I_b$, then stops when (S_0) or (S_1) are satisfied.

Although in general, in each on/off-state, we may have different behaviours, e.g. applying different techniques to reduce the (time dependent) transmission rate, for this work we assume that there is only one behaviour in the on-state. Such corresponds to restrict the values of the parameter ξ to the set $\{\beta_-, \beta_+\}$, for given constants $\beta_- < 0 < \beta_+$. The value β_+ accounts for the natural increasing effect of the disease (i.e. no agent intervention) and β_- accounts for the result of an agent action for controlling/reducing the transmission rate. Thus, in this model, $\xi \equiv \xi(t)$ turns now to be a piecewise function defined on $\overline{\mathcal{T}}_{t_0}$ with values on $\{\beta_-, \beta_+\}$, where the discontinuity instances are precisely the switching times generated by the application of one of the agent strategies $(S_0) - (S_2)$. Moreover, $\xi(t_0) = \beta_+$, $I(t_0) < I_b$, and the system may alternate (none or some bounded number of times) between the values β_- and β_+. Hence, it makes sense to define $\xi_- \subseteq \overline{\mathcal{T}}_{t_0}$ as the support where $\xi(t) = \beta_-$, $\xi_+ \subseteq \overline{\mathcal{T}}_{t_0}$ as the support where $\xi(t) = \beta_+$, $N_\xi \in \mathbb{N}_0$ the number of switchings, and $t_0, t_1, \ldots, t_{N_\xi}$, with $t_0 < t_1 < \cdots < t_{N_\xi} < \mathcal{T}$, the corresponding times of switchings.

Realistic constraints impose further that, in Eq. (2)(b), we assume $\beta_\xi(t) > 0$ on $\overline{\mathcal{T}}_{t_0} \equiv [0, \mathcal{T}]$ and $I(t) > 0$ on $\overline{\mathcal{T}}_{t_0}$ (i.e. in the time window there are always infected individuals), otherwise the problem is not interesting or meaningful. Additionally, for mathematical reasons, we require that, besides β_ξ being a continuous integrable bounded function on $\overline{\mathcal{T}}_{t_0}$, to be defined also on $\mathbb{R} \backslash \overline{\mathcal{T}}_{t_0}$. In particular, we will have the following structure

$$\beta_\xi(t) = \beta_\xi(t_0) + \xi \sum_{i=1}^{N_\xi} \left[\xi(t_{i-1}) \left(\min\{t_i, t\} - t_{i-1} \right) \chi_{(t_{i-1}, +\infty)}(t) \right], \qquad (3)$$

where $\chi_S(t)$ is the characteristic function of the set S. Therefore, there are positive constants β_* and β^*, such that $\beta_\xi(t) \in [\beta_*, \beta^*]$ for $t \in \mathbb{R}$.

Regarding the triggers values there are two situations: $I_b > I_s$ and $I_b < I_s$. The most natural situations is $I_b > I_s$, but $I_b < I_s$ makes sense in specific and limit situations. In either cases, because of the agent action, $\beta_\xi(t) \equiv \beta_\xi(t, I(t))$ and there is a memory effect, not present in Eq. (2), which controls in which state the system is running. In general, the model under study is neither an ordinary differential equation or a differential inclusion, but can be treated in the setting of (generic) hybrid systems, e.g. see [10, 20] for definitions and properties.

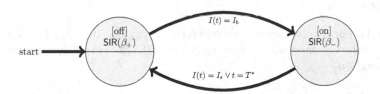

Fig. 1. Hybrid system associated to (2) describing the agent police.

The hybrid model is generally described in Fig. 1 and in more detail in Fig. 2, when expanding the invariant sets and dealing with the situations $I_b > I_s$ and $I_b < I_s$.

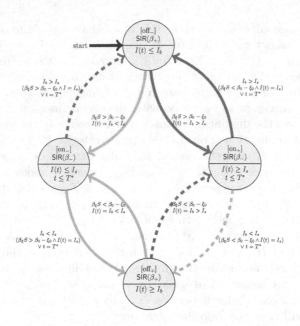

Fig. 2. Hybrid system associated to (2) with invariant sets; dashed edges mean the jumps are never used, since I_b, I_s are fixed parameters.

2.3 Existence and Uniqueness of Solutions

Let $x(t) = (S(t), I(t), R(t)) \in \Sigma_1$. The standard way to look to the system (2), as a dynamic process, is to consider it as the non-autonomous ODE Cauchy problem

$$\begin{cases} S'(t) = \alpha(1 - S(t)) + \zeta I(t) - \beta_\xi(t) I(t) S(t), \\ I'(t) = \beta_\xi(t) I(t) S(t) - (\zeta + \alpha + \gamma) I(t), \\ R(t) = 1 - S(t) - I(t), \end{cases} \qquad \Leftrightarrow \qquad \begin{cases} \dfrac{d\,x(t)}{dt} = F(t, x(t)), \\ x(0) = x_0 \in \Sigma_1, \end{cases}$$

where all parameters $\alpha, \zeta, \gamma, , \xi$ are fixed and, as described above, the initial conditions are values verifying $\beta_\xi(t_0) = \beta_0$, $0 \le S(t_0), I(t_0), R(t_0) \le 1$, $S(t_0) + I(t_0) + R(t_0) = 1$.

Lemma 1. *For any admissible parameters $\alpha, \zeta, \gamma, \beta_0, \beta_-, \beta_+, \xi \in \{\beta_-, \beta_+\}$, initial conditions above, and a strategy S_i, $i \in \{0, 1, 2\}$, the hybrid system has a unique solution.*

3   Stability and Bifurcation in Each Node

Restricting to a hybrid system node and for mathematical reasons, we may assume that the infectivity function β_ξ, satisfying (2)(b) on $[t_0, \mathcal{T}]$, is defined in all $\mathbb{R}$ with the structure

$$\beta_\xi(t) = \beta_0 \chi_{(-\infty, t_0]}(t) + (\beta_0 + \xi(t - t_0)) \chi_{(t_0, \mathcal{T}]}(t) + (\beta_0 + \xi(\mathcal{T} - t_0)) \chi_{(\mathcal{T}, +\infty)}(t), \quad (4)$$

where $\beta_0 = \beta_\lambda(t_0) > \max\{0, -\xi\mathcal{T}\}$ and $\chi_S(t)$ is the characteristic function of the set S. Therefore, there are positive constants β_* and β^*, such that $\beta_\lambda(t) \in [\beta_*, \beta^*]$ for $t \in \mathbb{R}$. In fact, $\beta_* = \min\{\beta_0, \beta_0 + \xi\mathcal{T}\}$ and $\beta^* = \max\{\beta_0, \beta_0 + \xi\mathcal{T}\}$. Further, each node is a nonlinear non-autonomous ordinary differential equation, so standard results of (autonomous) SIR-models do not apply. Such happen even for non-autonomous linear system as $x' = A(t)x$. For instance, the well-known fact that the origin is globally asymptotically stable if the real part of all eigenvalues of the matrix A are negative turn out to be wrong, as the following Nemytskii-Vinograd counterexample shows

$$\dot{x} = A(t)x, \quad \text{with} \quad A(t) = \begin{bmatrix} 1 - 4\cos(2t)^2 & 2 + 2\sin(4t) \\ -2 + 2\sin(4t) & 1 - 2\sin(2t)^2 \end{bmatrix},$$

which has the constant eigenvalues $\lambda_1 = -1$ and $\lambda_2 = -1$, but have the fundamental matrix $X(t) \equiv X(t, 0) = \begin{bmatrix} e^t \sin(2t) \\ e^t \cos(2t) \end{bmatrix}$, so solutions are unstable. To overcome these difficulties, we will use tools as the Chueshov's notion of non-autonomous equilibrium solution in a random dynamical system through a pull-back limit, and the notions of Bohl exponents and exponential dichotomy. We start by briefly collecting stability results for the autonomous SIR-model in the next section.

3.1   Stability of the Autonomous SIR-Model

The corresponding autonomous SIR-model of Eq. (2)(a), i.e. when $\xi = 0$, may be written as

$$\begin{cases} S'(t) = \alpha(1 - S(t)) + \zeta I(t) - \beta_0 I(t)S(t), \\ I'(t) = \beta_0 I(t)S(t) - \beta_0 \mathcal{R}_0^{-1} I(t), \\ R(t) + S(t) + I(t) = 1, \end{cases} \quad \text{with} \quad \mathcal{R}_0 = \frac{\beta_0}{\zeta + \alpha + \gamma} > 0, \quad (5)$$

where $\mathcal{R}_0$ is the so-called basic reproduction number. When $\mathcal{R}_0 \leq 1$, the disease-free equilibrium $(S^*, I^*) = (1, 0)$ is a globally asymptotic equilibrium point, proved by using Lyapunov-LaSalle function $V(S, I) = I$ and the LaSalle's Invariance Principle in the compact positively invariant set Σ_1. The endemic equilibrium

$$(\bar{S}, \bar{I}) = \left(\mathcal{R}_0^{-1}, \frac{\alpha}{\alpha + \gamma}(1 - \mathcal{R}_0^{-1}) \right) \quad (6)$$

only belongs to the simplex Σ_1 if $\mathcal{R}_0 > 1$. In fact, for $\mathcal{R}_0 > 1$, the disease-free equilibrium is unstable and the endemic equilibrium is asymptotically stable on $\Sigma_1 \backslash M_0$, where $M_0 = [0, 1] \times \{0\}$ is the stable manifold of the disease-free equilibrium, see [1]. The stability is obtained by using the Lyapunov function

$$V(S, I) = (S - \bar{S}) - \frac{\alpha + \gamma}{\beta_0} \log \frac{\beta_0 S - \zeta}{\beta_0 \bar{S} - \zeta} + (I - \bar{I}) - \bar{I} \log \frac{I}{\bar{I}}.$$

For the linear growth infectivity setting, i.e. $\xi \neq 0$, the disease-free equilibrium $(S^*(t), I^*(t)) = (1, 0)$ is still valid (in some regimes) but the endemic equilibrium $(\bar{S}, \bar{I})$ do not make sense a priori as an equilibrium point, since the ODE is non-autonomous. Hence, we introduce the notion of nontrivial non-autonomous attractor, obtained by a pullback limit mechanism when using the skew product flow formalism, in the next two sections.

3.2 The Skew Product Flow Formalism

From Lemma 1, for each initial condition $x_0 \in \Sigma_1$, there exists a unique solution $x(t; t_0, x_0)$, so we may define the flow $\phi_{t,t_0}(x_0) = x(t)$, which satisfies the following set of conditions:

(P_0) (a) $\phi_{t_0,t_0}(t_0) = x_0$; (b) $\phi_{t_2,t_0} = \phi_{t_2,t_1} \circ \phi_{t_1,t_0}$ for all $t_0 \le t_1 \le t_2$; (c) $t \mapsto \phi_{t,t_0}$ is differentiable and $(t, t_0, x_0) \mapsto \phi_{t,t_0}(x_0)$ is continuous.

Let (X, d_X) and (P, d_P) be metric spaces. A skew product flow $(, \varphi)$ is defined in terms of a cocycle mapping $\varphi : \mathbb{R}_0^+ \times P \times X \to X$ which is driven by an autonomous dynamical system $\psi : \mathbb{R} \times P \to P$ acting on a base or parameter space P and the time set $\mathbb{R}$. For convenience, we write $\varphi_t^p(x)$ to denote $\varphi(t, p, x)$. The driving system ψ on P is a group of homeomorphisms $(\psi)_{t \in \mathbb{R}}$ under the composition on P with the properties that

(P_1) (a) $\psi_0(p) = p$ for all $p \in P$; (b) $\psi_{s+t} = \psi_s \circ \psi_t$ for all $s, t \in \mathbb{R}$; (c) the mapping $p \mapsto \psi_t(p)$ is continuous;

(P_2) (a) $\varphi_0^p(x) = x$ for all $(p, x) \in P \times X$; (b) $\varphi_{t+s}^p = \varphi_t^q \circ \varphi_s^p$ with $q = \psi_s(p)$ for all $s, t \in \mathbb{R}_0^+$, and $p \in P$; (c) the mapping $(t, p, x) \mapsto \varphi(t, p, x)$ is continuous.

In particular, system (2) can be seen as

$$(a) \begin{cases} S' = \alpha(1 - S) + \zeta I - \beta_\xi I S, \\ I' = \beta_\xi I S - (\zeta + \alpha + \gamma) I, \\ R' = (\gamma I - \alpha R), \end{cases} \quad \text{and} \quad (b) \begin{cases} \beta'_\xi = \xi, \\ \xi' =' = \xi' = \alpha' = \zeta' = \gamma' = 0, \\ \tau' = 1. \end{cases} \quad (7)$$

So, we have $X = \Sigma_1$, $P = [\beta_*, \beta^*] \times \{\beta_-, \beta_+\} \times \{0, 1\} \times \{0, 1\} \times [0, \alpha^*] \times [0, \zeta^*] \times [0, \gamma^*] \times \{0\}$ for some $\alpha^*, \zeta^*, \gamma^* \in \mathbb{R}^+$, $\psi_t(p) = (p_0 + \xi t, p_1, p_2, p_3, p_4, p_5, p_6, p_7, p_8 + t)$ and Eq. (2)(a) are written as

$$\frac{d\,x(t)}{dt} = f(\psi_t(p), x(t)), \quad x(0) = x_0 \in X, \ p \in P. \quad (8)$$

One of the advantages of this formalism is that, in our case, X and P are both compact metric spaces. Natural extensions to random dynamical systems are obtained when replacing P by a probability space and the continuity property in $(P_1)(c)$ by measurability. The solutions are then generated by solutions of the corresponding Itô stochastic differential equation, see [8] as an introduction to the subject.

3.3   Non-autonomous Attractors

An entire solution is a continuous function $u : \mathbb{R} \to \Sigma_1$ such that $u(t + s) = \varphi_t^p(u(s))$ for all $s \in \mathbb{R}$ and $t \in \mathbb{R}_0^+$. We say that $\mathcal{A}$ is a non-autonomous attractor p-family if it is a set of nonempty compact subsets $A_p \subseteq \Sigma_1$ such that $\varphi_t^p(A_p) = A_{\psi_t(p)}$ for all $t \in \mathbb{R}_0^+$ and $p \in P$. Such sets are made up of entire solutions. Let $\dim_S(A, B)$ denote the Hausdorff semidistance between the nonempty compact subsets A and B of S. We call a non-autonomous attractor p-family $\mathcal{A}$ a pullback attractor p-family if it holds the pullback convergence

$$\lim_{t \to +\infty} \text{dist}_{\Sigma_1} \left(\varphi_t^{q_-}(D), A_p \right) = 0 \quad \text{with} \quad q_- = \psi_{-t}(p), \quad \forall p \in P, D \subseteq \Sigma_1, D \neq \emptyset,$$

and a forward attractor p-family if it holds the forward convergence

$$\lim_{t \to +\infty} \text{dist}_{\Sigma_1} \left(\varphi_t^p(D), A_{q_+} \right) = 0 \quad \text{with} \quad q_+ = \psi_t(p), \quad \forall p \in P, D \subseteq \Sigma_1, D \neq \emptyset.$$

If the convergence is uniform in p, the pullback and forward convergences coincide. A pullback absorbing set B is a nonempty subset of Σ_1 such that, for all $p \in P$ and (bounded) $D \subseteq \Sigma_1$, there exists a time $T \equiv T(p, D) > 0$, and

$$\varphi_t^q(D) \subset B \quad \text{with} \quad q = \psi_{-t}(p), \quad \forall t \geq T,$$

If there exists a pullback absorbing set B then $\mathcal{A}$ is a pullback attractor p-family if we define

$$A_p = \bigcap_{s > 0} \overline{\bigcup_{t > s} \varphi_t^{q_-}(B)}.$$

There is a corresponding formulation for t-families. Recall the system representation (2.3) with flow ϕ_{t,t_0}. We say that $\mathcal{A}$ is a non-autonomous attractor t-family if it is a set of nonempty compact subsets $A_t \subseteq \Sigma_1$ such that $\phi_{t,t_0}(A_{t_0}) = A_t$ for all $t \geq t_0$. Then it is a pullback attractor t-family if

$$\lim_{t_0 \to -\infty} \text{dist}_{\Sigma_1} \left(\phi_{t,t_0}(D), A_t \right) = 0, \quad \forall D \subseteq \Sigma_1, D \neq \emptyset. \tag{9}$$

This notion will play an important role in finding (nontrivial) non-autonomous equilibrium points.

3.4   SI(ξ) with $\xi \in \{\beta_-, \beta_+\}$

Considering the complexity of the model (2), in a first step, we study a sub-case proposed in [13], where the equation for $R(t)$ will not appear and the system evolves on the simplex

$$\Sigma_0 = \{(S, I) \in \mathbb{R}^2 : S, I \geq 0, \ S + I = 1\}.$$

Thus, it have the form

$$(a) \begin{cases} S' = \alpha(1 - S) + \zeta I - \beta_\xi IS, \\ I' = \beta_\xi IS - (\alpha + \zeta)I, \\ (S(t), I(t)) \in \Sigma_0, \end{cases} \quad (b) \ \beta'_\xi = \xi. \tag{10}$$

and will be denoted by SI(ξ)-model, although other variations exist on the literature with similar notation. For studying the stability properties of (10), it will be relevant the following results which may be checked computationally.

Lemma 2. *Assume $A > 0$ and $B \in \mathbb{R}$. If*

$$H_{\pm}[A, B](t, t_0) = \int_{t_0}^{t} e^{\pm(Ar + \frac{1}{2}Br^2)} dr,$$

then

$$H_{\pm}[A, B](t, t_0) = \begin{cases} \pm A^{-1}\left(e^{-At} - e^{-At_0}\right) & \text{if } B = 0, \\ \mp\sqrt{\mp\frac{\pi}{2B}} e^{\mp\frac{A^2}{2B}}\left(E_{A,B}(t) - E_{A,B}(t_0)\right) & \text{if } B \neq 0, \end{cases}$$

where erf is the error function and $E_{A,B}(t) = erf\left((A + Bt)\sqrt{\mp\frac{1}{2B}}\right)$.

Lemma 3. *The general Bernoulli differential equation*

$$x' = a(t)x - b(t)x^2, \quad x(t_0) = x_0,$$

for some arbitrary functions a and b, has the unique solution

$$x(t; t_0) = \frac{x_0\varphi(t, t_0)}{1 + x_0 \int_{t_0}^{t} b(s)\varphi(s, t_0)\, ds} \quad \text{with} \quad \varphi(s, s_0) = e^{\int_{s_0}^{s} a(r)\, dr}.$$

System (10)(a) can be reduced to the Bernoulli differential equation

$$I' = (\beta_\xi - \alpha - \zeta)I - \beta_\xi I^2, \quad I(t_0) = I_0 \in (0, 1), \tag{11}$$

with explicit solution (see Lemma 3)

$$I(t; t_0) = \frac{I_0\varphi(t, t_0)}{1 + I_0 \int_{t_0}^{t} \beta_\xi(s)\varphi(s, t_0)\, ds} \quad \text{and} \quad \varphi(s, s_0) = e^{\int_{s_0}^{s} \beta_\xi(r) - \alpha - \zeta\, dr}.$$

Let $\xi_0 = \beta_0 - \alpha - \zeta$. For the $t \mapsto \beta_\xi(t)$ function, $\xi_0 \in \mathbb{R}$ is the so-called shovel bifurcation parameter, due to a change in the range, and ξ is a transcritical bifurcation parameter, due to a change in the amplitude (see [12]).

Denoting by χ_C the characteristic function of the condition C, i.e. it is equal to one if C is true and zero otherwise, we have

$$\varphi(s, s_0) = \hat\varphi(s)\hat\varphi(s_0)^{-1} \quad \text{with} \quad \hat\varphi(r) = e^{\xi_0 r + \frac{1}{2}\xi r^2}\chi_{\{r \geq 0\}}, \tag{12}$$

so $\ln(\varphi(s, s_0)) = \xi_0(s - s_0)$, when $s_0 \leq s < 0$, and $\ln(\varphi(s, s_0)) = \xi_0(s - s_0) + \frac{1}{2}\xi\left(s^2 - s_0^2\right)$, when $s_0 \geq s \geq 0$. Then, the solution of the non-autonomous $SI(\xi)$-model has the explicit solution

$$I(t; t_0) = \left(1 + (I_0^{-1} - 1)e^{-\xi_0(t - t_0) - \frac{1}{2}\xi(t^2 - t_0^2)} + (\alpha + \zeta)e^{-\xi_0 t - \frac{1}{2}\xi t^2} H_+[\xi_0, \xi](t, t_0)\right)^{-1} \tag{13}$$

with $t_0 \geq 0, t \in \overline{T}_{t_0}$. In fact, from Lemma 3, we have

$$I(t; t_0) = \frac{I_0 e^{\xi_0(t-t_0)+\frac{1}{2}\xi(t^2-t_0^2)}}{1 + I_0 \int_{t_0}^t (\alpha + \zeta + \xi_0 + \xi s) e^{\xi_0(s-t_0)+\frac{1}{2}\xi(s^2-t_0^2)} \, ds}$$

with $t_0 \geq 0, t \in \overline{T}_{t_0}$. Further,

$$W = \int_{t_0}^t (\alpha + \zeta + \xi_0 + \xi s) e^{\xi_0(s-t_0)+\frac{1}{2}\xi(s^2-t_0^2)} \, ds$$

$$= (\alpha + \zeta) e^{-\xi_0 t_0 - \frac{1}{2}\xi t_0^2} \int_{t_0}^t e^{\xi_0 s + \frac{1}{2}\xi s^2} \, ds + e^{-\xi_0 t_0 - \frac{1}{2}\xi t_0^2} \int_{t_0}^t (\xi_0 + \xi s) e^{\xi_0 s + \frac{1}{2}\xi s^2} \, ds$$

$$= (\alpha + \zeta) e^{-\xi_0 t_0 - \frac{1}{2}\xi t_0^2} H_+[\xi_0, \xi](t, t_0) + e^{\xi_0(t-t_0)+\frac{1}{2}\xi(t^2-t_0^2)} - 1,$$

replacing on $I(t; t_0)$, we obtain closed form solution expression (13).

The Non-autonomous Equilibrium Solution. For simplicity of presentation and without loss of generality, let us assume that $t_0 = 0$. Following Chueshov [6], we consider the so-called non-autonomous equilibrium solution in a random dynamical systems set-up (E_t), found by taking the pullback limit $t_0 \to -\infty$ of (13) (see (9)), thus arriving to

$$I^*(t) = \left(\int_{-\infty}^t \beta_\xi(r) \varphi(t, r)^{-1} \, dr \right)^{-1}, \qquad S^*(t) = 1 - I^*(t).$$

Lemma 4. *Let* $t \in [0, T]$. *Define the complex value function*

$$G(t) \equiv H_+[\xi_0, \xi](t, 0) = \sqrt{-\frac{\pi}{2\xi}} e^{-\frac{\xi_0^2}{2\xi}} \left[\text{erf}\left((\xi_0 + \xi t)\sqrt{-\frac{1}{2\xi}} \right) - \text{erf}\left(\xi_0 \sqrt{-\frac{1}{2\xi}} \right) \right]$$

where erf is the error function. We have the following equilibrium E_t for SI(ξ):

(i) *When $\xi_0 \leq 0$, then $(S^*(t), I^*(t)) = (1, 0)$;*
(ii) *When $\xi_0 > 0$ and $\xi = 0$, then*

$$S^*(t) = (\alpha + \zeta)\beta_0^{-1} \qquad and \qquad I^*(t) = (\beta_0 - \alpha - \zeta)\beta_0^{-1};$$

(iii) *When $\xi_0 > 0$, $\xi \neq 0$ and $G(t) \in \mathbb{R}$, then*

$$S^*(t) = \frac{\xi_0^2 G(t) - (1 + \beta_0 G(t))\xi_0 + \beta_0}{\xi_0^2 G(t) - (1 + \beta_0 G(t))\xi_0 + \beta_0 + \xi_0 e^{\xi_0 t + \frac{1}{2}\xi t^2}}$$

and

$$I^*(t) = \frac{\xi_0 e^{\xi_0 t + \frac{1}{2}\xi t^2}}{\xi_0 e^{\xi_0 t + \frac{1}{2}\xi t^2} + \beta_0 - \xi_0[1 + (\beta - \xi_0)G(t)]}.$$

Moreover, the equilibrium E_t is globally asymptotically stable when $\xi_0 > \max\{0, -\xi T\}$.

The Disease-Free Equilibrium. The disease-free steady state equilibrium (E_0) is $(S^*(t), I^*(t)) = (1,0)$ on $\overline{\mathcal{T}}$. From above, E_t coincides with E_0 when $\xi \leq 0$. Kloeden-Kozyakin [13] (in Lemmas 4.1 and 3.1) proved a similar stability result, which we add for completeness and further reference.

Lemma 5. E_0 *is globally asymptotically stable w.r.t.* Σ_0, *when* $\beta^* \leq \alpha + \zeta$, *and unstable when* $\beta_* > \alpha + \zeta$.

Note that no information is given in the case $\beta_* \leq \alpha + \zeta < \beta^*$. A direct use of the above Lemma gives the following result.

Lemma 6. *The globally asymptotically stability of* E_0 *w.r.t.* Σ_0, *satisfies:*

(i) *If* $\xi = \beta_-$, *it is stable when* $\xi_0 \leq 0$ *and unstable when* $\xi_0 > \mathcal{T}|\beta_-|$;
(ii) *If* $\xi = \beta_+$, *it is stable when* $\xi_0 \leq -\mathcal{T}\beta_+$ *and unstable when* $\xi_0 > 0$.

We now define the so-called Bohl exponents, introduced by Bohl [5], which give information on the uniform exponential growth, whereas Lyapunov exponents, introduced by Lyapunov [17], only measure the exponential growth. It is well-known from ODEs that the Bohl exponent compared with the Lyapunov exponent is the appropriate concept in the setting of non-autonomous systems. For a summary of the history of Lyapunov and Bohl exponents see [7]. In detail, the upper Bohl exponent of a locally integrable function $f : \overline{\mathcal{T}} \to \mathbb{R}$ is defined by

$$\overline{B}_J(f) = \inf \left\{ w \in \mathbb{R} : \sup_{s \leq t, (s,t) \in J \times J} \frac{1}{t-s} \int_s^t f(r) - w\, dr < \infty \right\},$$

and the lower Bohl exponent by

$$\underline{B}_J(f) = \sup \left\{ w \in \mathbb{R} : \sup_{t \leq s, (s,t) \in J \times J} \frac{1}{t-s} \int_s^t f(r) - w\, dr < \infty \right\}.$$

The exponents are finite when f is integrally bounded, besides other properties (e.g. see [7]). From now on, we use the convection $a + [b, c] = [b + a, c + a]$ for any $a, b, c \in \mathbb{R}$.

We say that a linear system $x' = A(t)x$ has an exponential dichotomy (for short, E.D.) on $\mathbb{R}$, if there exists a projection $P : \mathbb{R}^n \to \mathbb{R}^n$ and positive constants C, α, β such that

$$\|\Phi(t)P\Phi^{-1}(s)\| \leq Ce^{-\alpha(t-s)}, \quad t \geq s,$$

$$\|\Phi(t)(\mathrm{Id} - P)\Phi^{-1}(s)\| \leq Ce^{-\beta(t-s)}, \quad s \geq t.$$

The dichotomy spectrum [22] is the set

$$\Sigma_A = \{c \in \mathbb{R} : x' = (A(t) - c\,\mathrm{Id})x \text{ admits no E.D.}\},$$

which is considered in the literature as the appropriate counterpart to eigenvalues in the non-autonomous setting. Then dichotomy spectrum is related with the Bohl exponents in the following way.

Lemma 7 (see [12]). *If* $f : \mathbb{R} \to \mathbb{R}$ *is a continuous bounded function and* $\lim_{t\to\pm\infty} f(t) = f_{\pm} \in \mathbb{R}$ *then* $\overline{\beta}_{\mathbb{R}}(f) = \max\{f_-, f_+\}$, $\underline{\beta}_{\mathbb{R}}(f) = \min\{f_-, f_+\}$, *and* $\overline{\beta}_{\mathbb{R}\pm}(f) = \underline{\beta}_{\mathbb{R}\pm}(f) = f_{\pm}$. *Moreover, if* $f : J \subseteq \mathbb{R} \to \mathbb{R}$ *is a continuous bounded function, then the dichotomy spectrum of* $\dot{x} = f(t)x$ *is given by* $[\underline{\beta}_J(f), \overline{\beta}_J(f)]$.

We have the following stability result.

Lemma 8. *The uniformly asymptotically stability of* E_0 *w.r.t.* Σ_0 *on* $\mathbb{R}$, *satisfies:*

(i) *If* $\xi = \beta_-$, *it is stable when* $\xi_0 \leq 0$ *and unstable when* $\xi_0 > T|\beta_-|$;
(ii) *If* $\xi = \beta_+$, *it is stable when* $\xi_0 \leq -T\beta_+$ *and unstable when* $\xi_0 > 0$.

E_0 *is uniformly asymptotically stable w.r.t.* Σ_0 *on* $\overline{T}_0$, *when* $\xi_0 \leq -\frac{1}{2}\xi T$, *and unstable when* $\xi_0 > -\frac{1}{2}\xi T$.

3.5 SIR(ξ) with $\xi \in \{\beta_-, \beta_+\}$

For simplicity, assume $t_0 = 0$. A basic observation regarding SIR(ξ) is that the sign of I' is, almost everywhere, given by $\nu(t) := \text{sign}(I') = \text{sign}((\beta_0 + \xi t) S - q) \in \{-1, +1\}$, so, $\nu(t)S(t) > \nu(t)\frac{q}{\beta_0 + \xi t}$ implies $\nu(t)I$ increases, where $q = \alpha + \zeta + \gamma$. Hence, defining the auxiliary functions

$$\psi_I(t) = \beta_\xi(t)S(t) - (\alpha + \zeta + \gamma) \quad \text{and} \quad \psi_R(t) = (\gamma - \alpha R(t)I(t)^{-1}),$$

and, since $I(t) > 0$ and $(\psi_I + \psi_R)I = \beta_\xi SI - \zeta I - \alpha(I+R) = \beta_\xi SI - \zeta I - \alpha(1-S)$, we have that SIR($\xi$) can be written as

$$(a) \begin{cases} S' = -(\psi_I + \psi_R)I, \\ I' = \psi_I I, \\ R' = \psi_R I, \end{cases} \quad \text{and} \quad (b) \ \beta'_\xi = \xi, \tag{14}$$

so the monotony of (S, I, R) are determined by the signs of $(-\psi_I - \psi_R, \psi_I, \psi_R)$.

Lemma 9. *For any* α, γ *and* $\beta_\xi(0)$, *there exist* $S(0), I(0), R(0) \in \Sigma_1$ *and (a small enough)* $T > 0$ *such that we may prescribe an arbitrary combination of monotonicity for* $S(t), I(t), R(t)$ *in* $t \in [0, T]$, *as solutions of* SIR(ξ).

Lemma 9 is false for SI(ξ), because $\text{sign}(S') = -\text{sign}(I')$, and also shows that the flow, associated with the hybrid system of Fig. 2, can be quite complex since each node can (generally) initiate in any monotonicity situation.

Poincaré, in 1892, started the theory of normal forms as a technique to simplifying a nonlinear system in the neighborhood of a reference solution by a smooth change of coordinates. Let us summarize the ideas. Consider the autonomous system $x' = Ax + f(x)$, where A is a constant matrix and $f(x) = O(\|x\|^2)$ as $\|x\| \to 0$. Then by a formal coordinate transformation $x = y + \sum_{i=2}^{+\infty} h_i y^i$ the above system can be changed into the system $y' = By + \sum_{i=1}^{+\infty} g_i y^i$ where B us the complex Jordan form of A, $g_i = (g_i^1, \ldots, g_i^n)$ and $g_i^j = 0$ if $\sum_{i=1}^{n} p_i \xi_i - \xi_j \neq 0$

for ξ_i eigenvalues of A, $p \in \mathbb{Z}_+^n$ and $\sum_{i=1}^n p_i \geq 2$. In addition, if f is analytic in the origin, we have the so-called Poincaré-Dulac's analytic normal forms. Consider the systems

$$(a) \ x' = A(t)x + f(t,x), \quad (b) \ x' = A(t)x. \tag{15}$$

A change of variables $x = P(t)y$ is said to be a Lyapunov-Perron transformation (for short, L.P.) if $P(t)$ is nonsingular for all $t \in \mathbb{R}$ and P, P^{-1}, P' are uniform bounded in $t \in \mathbb{R}$. The system $(15)(a)$ is locally analytically equivalent to the system $y' = G(t,y)$ if there exists a coordinate substitution $x = P(t)y + h(t,y)$ which transforms one to the other, where f, G, P, h are analytic in $\bar{B}_\rho(0) \times \mathbb{R}$, for some $\rho > 0$, $f(t,0) = G(t,0) = h(t,0) = 0$, P is a LP transformation and $h(t,y) = O(\|y\|^2)$ as $\|y\| \to 0$. Assume the dichotomy spectrum of $(15)(b)$ to be $\Sigma_A = [a_1, b_1] \cup \cdots \cup [a_p, b_p]$ where $a_1 \leq b_1 < \cdots < a_p \leq b_p$. We say that system $(15)(b)$ is of type: (type-I) when $a_1 b_p > 0$ (i.e. it is in the Poincaré domain); (type-II) when $a_1 b_p > 0$ and it is non-resonant, i.e.

$$0 \notin \left[\sum_{i=1}^p a_i m_i - a_j, \sum_{i=1}^p b_i m_i - b_j \right] \quad \text{with} \quad m \in \mathbb{N}^p;$$

(type-III) when $a_1 b_p > 0$ and $A(t)$ is block diagonal w.r.t. the spectral interval $[a_i, b_i]$ (i.e. of Poincaré-Dulac type).

Lemma 10. *(see [24]) We have for type-II, system $(15)(a)$ is locally analytically equivalent to its linear part $(15)(b)$.*

In this section, we always assume $x = (S, I, R)$, and $\gamma > 0$. Under the linear transformation

$$x = \begin{pmatrix} 1 - \frac{\gamma}{\zeta+\gamma} & \frac{\zeta}{\zeta+\gamma} \\ 0 & 0 & 1 \\ 0 & \frac{\gamma}{\zeta+\gamma} & -\frac{\gamma}{\zeta+\gamma} \end{pmatrix} z \quad \text{with} \quad \text{Det} = -\frac{\gamma}{\zeta + \gamma} \neq 0,$$

the system $(2)(a)$ is transformed into (the Jordan canonical linear form)

$$z' = \begin{pmatrix} 0 & 0 & 0 \\ 0 & -\alpha & 0 \\ 0 & 0 & -\alpha - \zeta - \gamma \end{pmatrix} z + (\gamma + \zeta)^{-1} z_3 (z_1(\gamma + \zeta) - z_2\gamma - z_3\zeta)\beta_\xi(t) \begin{pmatrix} 0 \\ 1 \\ 1 \end{pmatrix}.$$

This means that $z_1(t) = S(0) + I(0) + R(0) = 1$ is constant. Hence, for $(y_1, y_2) = (z_2, z_3)$, we have

$$y' = \begin{pmatrix} -\alpha & \beta_\xi(t) \\ 0 & \beta_\xi(t) - \alpha - \zeta - \gamma \end{pmatrix} y - (\gamma + \zeta)^{-1} \beta_\xi \begin{pmatrix} y_1 y_2 \gamma + y_2^2 \zeta \\ y_1 y_2 \gamma + y_2^2 \zeta \end{pmatrix}.$$

For $\mathsf{SIR}(\xi)$, recall that the equilibrium E_0 is $(S^*(t), I^*(t), R^*(t)) = (1, 0, 0)$. Hence, the equilibrium point $(y_1^*, y_2^*) = (0, 0)$ corresponds to E_0. Now, consider the linear transformation

$$y = \begin{pmatrix} -\frac{\beta_\xi(t)}{\beta_\xi(t)-\zeta-\gamma} & \frac{\beta_\xi(t)}{\beta_\xi(t)-\zeta-\gamma} \\ 0 & 1 \end{pmatrix} w \quad \text{with} \quad \text{Det} = -\frac{\beta_\xi(t)}{\beta_\xi(t) - \zeta - \gamma} \neq 0,$$

so

$$w_1(t) = -\frac{\beta_\xi - \zeta - \gamma}{\beta_\xi}\left(I(t) + \frac{\zeta + \gamma}{\gamma}R(t)\right) \quad \text{and} \quad w_2(t) = I(t), \qquad (16)$$

then we have the new system

$$w' = \begin{pmatrix} -\alpha & 0 \\ 0 & \xi_0 + \xi t \end{pmatrix} w + (\gamma + \zeta)^{-1}(\beta_\xi(t) - \zeta - \gamma)^{-1}h[w_1, w_2]\begin{pmatrix} \gamma + \zeta \\ \beta_\xi(t) \end{pmatrix}. \qquad (17)$$

where $\xi_0 = \beta_0 - \alpha - \zeta - \gamma$ and $h[w_1, w_2](t) = \gamma\beta_\xi(t)w_1w_2 + (\zeta\gamma - (\gamma + \zeta)\beta_\xi(t) + \zeta^2)w_2w_2$.

Lemma 11. *When $\xi_0 < -\frac{1}{2}\xi T$ and $\gamma \neq -\zeta$, $\mathsf{SIR}(\xi)$ is locally analytically equivalent to the linear system*

$$v' = \begin{pmatrix} 0 & 0 & 0 \\ 0 & -\alpha & 0 \\ 0 & 0 & \xi_0 + \xi t \end{pmatrix} v.$$

4   Three Illustrative Examples

We present numerical examples that illustrate occurrences that may not appear in SIR-models without agent actions: (E_1) the agent action is not able to decrease the number of infected individuals and they tend to a non-autonomous attractor (see Sect. 3.3 for a precise definition); (E_2) the agent action introduces an oscillatory behaviour in the number of infected individuals around some non-autonomous attractor; and (E_3) the agent action are able, in each period, to significantly decrease the number of infected individuals but such mechanism introduces a succession of bumps along time.

Since hybrid systems return solutions selected by discrete jump events its validation and error control is a key issue, requiring tailored tools based on first order logics. The numerical calculations of the given examples below were produced using two packages dReal and dReach [9,14,15]. dReal is an automated reasoning tool focused on solving problems that can be encoded as first-order logic formulas over the real numbers by implementing the framework of δ-complete decision procedures. dReach deals with the bounded δ-reachability problem. For a hybrid system $H = < X, Q, flow, jump, inv, init >$, where flow, jump, inv, init are SMT formulas that dReal can handle and specifying a numerical error bound δ, any formula ϕ can have its δ-perturbation counterpart ϕ^δ. Then, a δ-perturbation of H is defined as $H^\delta = < X, Q, flow^\delta, jump^\delta, inv^\delta, init^\delta >$, by relaxing the logic formulas in H. Now, choosing $n \in \mathbb{N}$ to be a bound on the number of discrete mode changes, $T \in \mathbb{R}^+$ an upper bound on the time duration, and *unsafe* to encode a subset of $X \times Q$, the bounded δ-reachability problem asks for one of the following answers: (a) "safe" if H cannot reach *unsafe* in n steps within time T; (b) "δ-unsafe" if H^δ can reach $unsafe^\delta$ in n steps within time T. In this way, we ensure that our examples are numerically correct, since the SMT tool produce a logical proof of reachability.

4.1   Example (E_1) – Nontrivial Asymptotically Stable Attractor

Figure 3 shows a numerical example where a nontrivial asymptotically stable attractor on $I(t)$ appears; the parameters are $I_b = 0.600$, $I_s = 0.500$, $T^* = +\infty$, $\beta_+ = 0.200$, $\beta_- = -1.300$, $\alpha = \zeta = \gamma = 0.100$, and the initial conditions $S(0) = 0.550$, $I(0) = 0.300$, $R(0) = 0.150$, $\beta(0) = 1.400$.

Although, there is no standard basic reproduction number in the non-autonomous setting (but there exist generalizations as the notion [2]), we would still look to the value of $\mathcal{R}_0(t) = \frac{\beta_\xi(t)}{\zeta + \alpha + \gamma}$. Further, when $\xi = \beta_+$, $\lim_{t \to +\infty} \mathcal{R}_0^{-1}(t) = 0$. Since $\mathcal{R}_0(t) \leq 1$ means $\beta_\xi(t) \leq 0.3$, it is expected that, for $4.5165 \leq t \leq 5.2294$, the solution is attracted by the disease-free equilibrium (S^*, I^*), and, for $0 \leq t < 4.5165$ and $t > 5.2294$, the solution is attracted by some (non-autonomous) endemic equilibrium which in the limit tends to $(\bar{S}, \bar{I}) = \left(0, \frac{\alpha}{\alpha + \gamma}\right) = (0, 0.5)$. In the figures, the change of color means transition between hybrid nodes/modes.

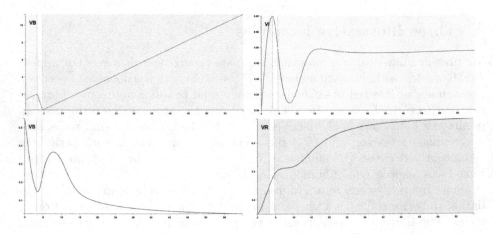

Fig. 3. A nontrivial asymptotically stable attractor for $I(t)$

4.2   Example (E_2) – Oscillatory Behaviour

Figure 4 shows a set of parameters for which the infected individuals variable $I(t)$ oscillates around some (non-autonomous) endemic equilibrium which in the limit tends to $(\bar{S}, \bar{I}) = \left(0, \frac{\alpha}{\alpha + \gamma}\right) = (0, 0.3)$; the parameters are $I_b = 0.300$, $I_s = 0.285$, $T^* = +\infty$, $\beta_+ = 0.200$, $\beta_- = -0.200$, $\alpha = \gamma = 0.100$, $\zeta = 0.200$, and the initial conditions $S(0) = 0.910$, $I(0) = 0.060$, $R(0) = 0.030$, $\beta(0) = 0.400$.

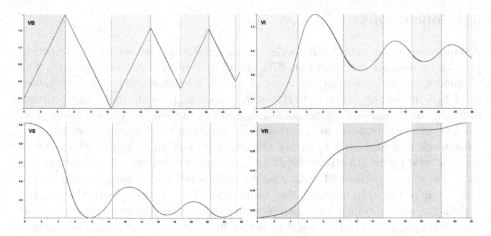

Fig. 4. Oscillatory behaviour of $I(t)$

4.3   Example (E_3) – Bump Behaviour

Figure 5 shows the most interesting profile for our model since a succession of bump behaviours appear along time, although in the intervals between bumps $I(t)$ is coming near to zero; the parameters are $I_b = 0.100$, $I_s = 0.050$, $T^* = +\infty$, $\beta_+ = 0.200$, $\beta_- = -1.300$, $\alpha = \zeta = \gamma = 0.100$, and the initial conditions $S(0) = 0.925$, $I(0) = 0.050$, $R(0) = 0.025$, $\beta(0) = 1.400$.

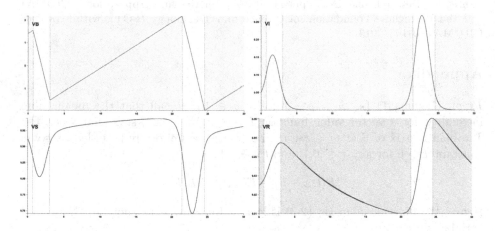

Fig. 5. Bump behaviour of $I(t)$

5   Conclusions

A non-autonomous hybrid SIR-model was introduced as the result of agent action policies on diseases modelled by a SIR-model with linear infectivity growth. The coupled system shows a great variety of profile solutions and extends the standard SIR-model (i.e. when $\max I(t) < I_b$). Two ingredients make the problem difficult: (a) its non-autonomous nature; and (b) the jumps between ODEs (i.e. the hybrid system nodes) are controlled by the values of the state variable $I(t)$. This work is a first step to study the properties of such hybrid SIR-models, since several issues are still to be clear, e.g. complete scheme of stability of the hybrid system, existence of nontrivial periodic solutions crossing several nodes, behaviour of the system under the assumption $I_s > I_b$, etc. Nevertheless, the stability results obtained and examples provided already show the richness and potential application of the model to better fit oscillatory real data. Considering that the number of infected individuals is the most observable variable in reality, Fig. 5 turns out to be quite interesting, since the choices of I_b, I_s are critical to determine the solution profile. Hence, it raises several questions: (a) Are human disease control strategies somehow responsible for the oscillatory behaviour of some diseases? (b) For each choice of parameters and agent action, are there optimal values for I_b, I_s such the maximum of $I(t)$ is reduced? (c) How different is the solution if the agent action activates by a stochastic process?

Acknowledgements. This research was partial supported by Portuguese funds through the FCT project "DALI – Dynamic logics for cyber-physical systems: towards contract based design" with reference P2020-PTDC/EEI-CTP/4836/2014; the Center for Research and Development in Mathematics and Applications (CIDMA) and the Portuguese Foundation for Science and Technology (FCT), within project UID/MAT/04106/2013.

Appendix

Proof. (Lemma 1) Let $h = y - x$ for $x, y \in \Sigma_1$. Recall that the mean value inequality, for a vector value function $F : \mathbb{R} \times \Sigma_1 \to \mathbb{R}^3$, says that when the Jacobian matrix of F at $w = x + \tau h$, i.e. $J_F(w)$, is uniformly bounded by some constant $L > 0$ for any $\tau \in [0,1]$ and $t \in \mathbb{R}$, then

$$|F(t, x + h) - F(t, x)| \leq L|h|.$$

Hence, the function F in Eq. (2.3) is locally Lipschitz continuous in the second variable since

$$J_F(w(t)) = \begin{pmatrix} -\alpha - \beta_\xi(t)w_2(t) & \zeta - \beta_\xi(t)w_1(t) & 0 \\ \beta_\xi(t)w_2(t) & \beta_\xi(t)w_1(t) - (\zeta + \alpha + \gamma) & 0 \\ 0 & \gamma & -\alpha \end{pmatrix}$$

is uniformly bounded, because $\beta_\xi(t) \in [\beta_*, \beta^*]$ and $w(t)$ is bounded for any $\tau \in [0,1]$, by definition. So, the Picard-Lindelöf theorem ensures the existence and uniqueness of solution in each node of the hybrid system. Its not difficult to see that the solution is globally defined. Further, the hybrid system Fig. 2 is deterministic and has only one jump condition in each node, so we conclude the proof.

Proof. (Lemma 4) First note that

$$I^*(t)^{-1} = \int_{-\infty}^t \beta_\xi(r)\varphi(t,r)^{-1}\,dr = \hat{\varphi}(t)^{-1}\lim_{a\to+\infty}\int_{-a}^t \beta_\xi(r)\hat{\varphi}(r)dr$$
$$= \hat{\varphi}(t)^{-1}\int_0^t \beta_\xi(r)\hat{\varphi}(r)dr + \hat{\varphi}(t)^{-1}\lim_{a\to+\infty}\int_{-a}^0 \beta_\xi(r)\hat{\varphi}(r)dr$$

with $t \in [0,\mathcal{T}]$. We consider two cases: (a) $\xi = 0$; and (b) $\xi \geq 0$.

(a) For $\xi = 0$, i.e. when $\beta_\xi(t) \equiv \beta_0$ is constant and $\hat{\varphi}(r) = e^{\xi_0 r}$, we have

$$I^*(t)^{-1} = \beta_0 e^{-\xi_0 t}\int_0^t e^{\xi_0 r}dr + \beta_0 e^{-\xi_0 t}\lim_{a\to+\infty}\int_{-a}^0 e^{\xi_0 r}dr$$
$$= \frac{\beta_0}{\xi_0}\left(1 - \lim_{a\to+\infty}e^{-\xi_0 a}\right).$$

From which, we obtain: (i) for $\xi_0 \leq 0$, then $(S^*(t), I^*(t)) = (1,0)$; and (b) for $\xi_0 > 0$, we recover the expected values

$$I^*(t) = (\beta_0 - \alpha - \zeta)\beta_0^{-1} \quad\text{and}\quad S^*(t) = (\alpha + \zeta)\beta_0^{-1}. \tag{18}$$

(b) In general, for $\xi \neq 0$ and $t \in \mathcal{T}_{t_0}$, using (12) and integration by parts, we have

$$I^*(t)^{-1}\hat{\varphi}(t) = \int_0^t (\beta_0 + \xi r)e^{\xi_0 r + \frac{1}{2}\xi r^2}\,dr + \beta_0 \lim_{a\to+\infty}\int_{-a}^0 e^{\xi_0 r}dr$$
$$= \int_0^t (\beta_0 + \xi r)e^{\xi_0 r + \frac{1}{2}\xi r^2}\,dr + \lim_{a\to+\infty}\frac{\beta_0}{\xi_0}\left(1 - e^{-\xi_0 a}\right)$$
$$= \hat{\varphi}(t) - 1 - (\alpha + \zeta)G(t) + \lim_{a\to+\infty}\frac{\beta_0}{\xi_0}\left(1 - e^{-\xi_0 a}\right).$$

Therefore, we obtain: (i) for $\xi_0 \leq 0$, then $(S^*(t), I^*(t)) = (1,0)$; and (ii) for $\zeta_0 > 0$, we get

$$I^*(t) = \frac{\beta_0 - \alpha - \zeta}{\xi_0 + \beta_0\hat{\varphi}(t)^{-1} - \xi_0[1 + (\alpha + \zeta)G(t)]\hat{\varphi}(t)^{-1}} \quad\text{and}\quad S^*(t) = 1 - I^*(t).$$
$$\tag{19}$$

In particular, when $\xi \to 0$, $G(t) \to (1 - \hat{\varphi}(t))\xi_0^{-1}$ so we recover the values (18).

To prove the stability result, assume $\xi_0 > \max\{0, -\xi\mathcal{T}\}$. This means that $\beta_* > \alpha + \zeta$. For any solution $I(t)$, $D = I(t) - I^*(t)$ satisfy

$$D'(t) = (\beta_\xi(t) - \alpha - \zeta)D(t) \geq (\beta_* - \alpha - \zeta)D(t) \quad\Rightarrow\quad D'(t)^2 \leq D(0)^2\, e^{-2|\beta_* - \alpha - \zeta|t},$$

so $D(t) \to 0$ and $I(t) - I^*(t) \to 0$ as $t \to +\infty$. Because $|S(t) - S^*(t)| = |1 - I(t) - (1 - I^*(t))| = |I^*(t) - I(t)|$, we have the desired conclusion.

Proof. (Lemma 5) Recall $\beta_\xi(t) \in [\beta_*, \beta^*]$ for $t \in \mathbb{R}$. We consider three cases: (A) $\beta^* < \alpha + \zeta$, (B) $\beta^* = \alpha + \zeta$, and (C) $\beta_* > \alpha + \zeta$. (A) Using the second equation of (10) and $0 \leq S \leq 1$, we obtain $(I^2)' = 2I\,I' = 2\,(\beta_\xi S - \alpha - \zeta)\,I^2 \leq 2\,(\beta^* - \alpha - \zeta)\,I^2$ which implies that $I(t)^2 \leq I(0)^2 e^{-2|\beta^* - \alpha - \zeta|t}$ as $t \to +\infty$, meaning that E_0 is (asymptotically) stable. (B) Consider Eq. (11), so for all $I > 0$, we have $I' = (\beta_\xi - \alpha - \zeta)I - \beta_\xi I^2 = -\beta_\xi I^2 < -\frac{\beta_\xi}{\beta_*}I^2 < 0$ and $0 \leq I(t) \leq \lim_{t\to+\infty} \frac{\beta_* I(0)}{\beta_* + \beta_\xi I(0)t}$. Thus, $S(t) - S^*(t) = S(t) - 1 = -I(t) \to 0$ as $t \to +\infty$. Hence, E_0 is (asymptotically) stable. (C) Suppose $I(t) \leq \epsilon \in [0,1]$, so $S(t) \geq 1 - \epsilon$. Then

$$S' \leq [\alpha + \zeta - \beta_\xi(1 - \epsilon)]\,I \quad \text{and} \quad I' \geq [\beta_*(1 - \epsilon) - \alpha - \zeta]\,I,$$

so I is strictly increasing if $0 < I(t) \leq \epsilon$ and $\epsilon < 1 - (\alpha + \zeta)\beta_*^{-1}$. In particular, for any solution if $(S(t), I(t)) \to (1, 0)$ then we have that $I(t)$ is strictly increasing (i.e. a contradiction), so E_0 is unstable.

Proof. (Lemma 6) Recall that $\beta_0 > \max\{0, -\xi\mathcal{T}\}$, $\beta_* = \min\{\beta_0, \beta_0 + \xi\mathcal{T}\}$ and $\beta^* = \max\{\beta_0, \beta_0 + \xi\mathcal{T}\}$. So, by direct computation, we get

$$\max\{\beta_0, \beta_0 + \xi\mathcal{T}\} \leq \alpha + \zeta \Leftrightarrow \max\{0, \xi\mathcal{T}\} \leq -\xi_0 \Leftrightarrow \xi_0 \leq \min\{0, -\xi\mathcal{T}\},$$

$$\min\{\beta_0, \beta_0 + \xi\mathcal{T}\} > \alpha + \zeta \Leftrightarrow \min\{0, \xi\mathcal{T}\} > -\xi_0 \Leftrightarrow \xi_0 > \max\{0, -\xi\mathcal{T}\},$$

for which Lemma 5 implies that E_0 is globally asymptotically stable w.r.t. Σ_0, when $\xi_0 \leq \min\{0, -\xi\mathcal{T}\}$, and unstable when $\xi_0 > \max\{0, -\xi\mathcal{T}\}$. The statements in this lemma are then a direct consequence of $\mathcal{T} \geq 0$ and $\beta_- < 0 < \beta_+$.

Proof. (Lemma 8) First, suppose the system (10) is defined on $t \in J \subseteq \mathbb{R}$. By applying Lemma 7, we conclude that the linear part of (10) has the dichotomy spectrum $-(\alpha + \zeta) + [\underline{\beta}_J(\beta_\xi), \overline{\beta}_J(\beta_\xi)]$. Hence, by propositions 4.9 and 4.10 in [19], Lemma 5 is still valid when we replace *globally asymptotically stable* by *uniformly asymptotically stable* and β_*, β^* by $\underline{\beta}_J(\beta_\xi), \overline{\beta}_J(\beta_\xi)$, respectively. This tell us that it is expect to occur a bifurcation of (10) when $\underline{\beta}_J(\beta_\xi) = \overline{\beta}_J(\beta_\xi) = \alpha + \zeta$.

We have $\underline{\beta}_{\overline{\mathcal{T}}_0}(\beta_\xi) = \overline{\beta}_{\overline{\mathcal{T}}_0}(\beta_\xi) = \beta_0 + \frac{1}{2}\xi\mathcal{T}$. In fact, note that β_ξ is a integrable bounded function in $\overline{\mathcal{T}}_0$. For $s, t \in \overline{\mathcal{T}}_0$ and $w \in \mathbb{R}$, let

$$F(s, t, w) = \frac{1}{t - s}\int_s^t \beta_\xi(r) - w\,dr = \beta_0 + \frac{1}{2}\xi(t + s) - w,$$

$$\sup_{s \leq t, (s,t) \in [0,\mathcal{T}]^2} F(s, t, w) = \sup_{t \leq s, (s,t) \in [0,\mathcal{T}]^2} F(s, t, w) = \beta_0 + \frac{1}{2}\xi\mathcal{T} - w.$$

Hence, $\underline{\beta}_{\overline{\mathcal{T}}_0}(\beta_\xi) = \overline{\beta}_{\overline{\mathcal{T}}_0}(\beta_\xi) = \beta_0 + \frac{1}{2}\xi\mathcal{T}$. So, replacing the obtained values in $\overline{\beta}_{\overline{\mathcal{T}}_0}(\beta_\xi) \leq \alpha + \zeta$, $\underline{\beta}_{\overline{\mathcal{T}}_0}(\beta_\xi) > \alpha + \zeta$ and recalling that $\xi_0 = \beta_0 - \alpha - \zeta$, we simplify to

$$\beta_0 + \frac{1}{2}\xi\mathcal{T} \leq \alpha + \zeta \Leftrightarrow \xi_0 \leq -\frac{1}{2}\xi\mathcal{T} \quad \text{and} \quad \beta_0 + \frac{1}{2}\xi\mathcal{T} > \alpha + \zeta \Leftrightarrow \xi_0 > -\frac{1}{2}\xi\mathcal{T},$$

which confirms the above bifurcation point of (10).

By Lemma 7, we have that $\underline{\beta}_{\mathbb{R}}(\beta_\xi) = \min\{\beta_0, \beta_0 + \xi T\}$ and $\overline{\beta}_{\mathbb{R}}(\beta_\xi) = \max\{\beta_0, \beta_0 + \xi T\}$. In the same way, for $t \in \mathbb{R}$, we have

$$\max\{\beta_0, \beta_0 + \xi T\} \leq \alpha + \zeta \Leftrightarrow \xi_0 \leq \min\{0, -\xi T\},$$

$$\min\{\beta_0, \beta_0 + \xi T\} > \alpha + \zeta \Leftrightarrow \xi_0 > \max\{0, -\xi T\},$$

The statement in the lemma is a consequence of $T \geq 0$ and $\beta_- < 0 < \beta_+$.

Proof. (Lemma 9) Let $v_1 = \psi_I(0)$ and $v_2 = \psi_R(0)$. The functions ψ_I, ψ_R are continuous so there exist $T > 0$ such that their signs are preserved in $[0, T]$, so from (14) they prescribe the monotonicity of $S(t), I(t), R(t)$ in $t \in [0, T]$. Now, it is enough to explicitly construct the map $(v_1, v_2) \mapsto (S(0), I(0), R(0))$ as

$$S(0) = \frac{v_1 + q}{\beta_0}, \quad I(0) = -\frac{\alpha(-\beta_0 + v_1 + q)}{\beta_0(\alpha + \gamma - v_2)},$$

$$R(0) = -\frac{\gamma^2 + (v_1 - \beta_0 - v_2 + \alpha + \zeta)\gamma + v_2(\beta_0 - v_1 - \alpha - \zeta)}{\beta_0(\alpha + \gamma - v_2)},$$

where $q = \alpha + \zeta + \gamma$, $\beta_0 = \beta_\xi(0)$ and $v_1 \in (-q, -q + \beta_0)$, $v_2 \in (\gamma, \gamma + \alpha)$ (which ensure $0 < S(0), I(0), R(0) < 1$).

Proof. (Lemma 11) Consider the (reduced) system (17). The dichotomy spectrum (of the linear part) is

$$\Sigma_A = [-\alpha, -\alpha] \cup [\underline{\beta}_{\overline{T}}(\beta_\xi) - \alpha - \zeta - \gamma, \overline{\beta}_{\overline{T}}(\beta_\xi) - \alpha - \zeta - \gamma].$$

Recall $\underline{\beta}_{\overline{T}_0}(\beta_\xi) = \overline{\beta}_{\overline{T}_0}(\beta_\xi) = \beta_0 + \frac{1}{2}\xi T$, First, suppose $\xi_0 > -\alpha - \frac{1}{2}\xi T$ so we have the (ordered) dichotomy spectrum

$$\Sigma_A = [-\alpha, -\alpha] \cup \left[\xi_0 + \frac{1}{2}\xi T, \xi_0 + \frac{1}{2}\xi T\right]$$

and $a_1 b_2 = -\alpha\left(\xi_0 + \frac{1}{2}\xi T\right) > 0$, meaning it is a system of type-I. Since $a_i = b_i$, we also have that, for $m \in \mathbb{N}^2$, to be of type-II is the same as

$$0 \notin \left\{-\alpha m_1 + \left(\xi_0 + \frac{1}{2}\xi T\right) m_2 + \alpha, -\alpha m_1 + \left(\xi_0 + \frac{1}{2}\xi T\right) m_2 - \xi_0 - \frac{1}{2}\xi T\right\},$$

Which is true, since the inclusion $am_1 + bm_2 \in \{a, b\}$ with $a = \alpha$, $b = |\xi_0 + \frac{1}{2}\xi T|$, do not have integer solutions. Hence, it is of type-II. If we suppose $\xi_0 < -\alpha - \frac{1}{2}\xi T$ so we have the (ordered) dichotomy spectrum

$$\Sigma_A = \left[\xi_0 + \frac{1}{2}\xi T, \xi_0 + \frac{1}{2}\xi T\right] \cup [-\alpha, -\alpha],$$

the conclusions are the same.

Then system (17) is locally analytically equivalent to its linear part $w' = A(t)w$, by applying Lemma 10. From (16), there exists a matrix $Q(t)$, with determinant $\gamma^{-1}\beta_\xi^{-1}(\beta_\xi - \zeta - \gamma)(\zeta + \gamma)$, such that $w = Q(t)x$, so $x' = Q^{-1}(t)A(t)Q(t)x$ and then applying the Jordan canonical form transformation $x = Bv$, i.e.

$$x' = \begin{pmatrix} 0 & 0 & 0 \\ 0 & \xi_0 + \xi t & 0 \\ 0 & -\gamma(\zeta + \gamma)^{-1}(\alpha + \xi_0 + \xi t) & -\alpha \end{pmatrix} x \quad \text{and} \quad B = \begin{pmatrix} 1 & 0 & 0 \\ 0 & 0 & 1 \\ 0 & \frac{\gamma}{\zeta + \gamma} & -\frac{\gamma}{\zeta + \gamma} \end{pmatrix},$$

gives the expected result.

References

1. Adda, P., Bichara, D.: Global stability for SIR and SIRS models with differential mortality. Int. J. Pure Appl. Math. **80**(3), 425–433 (2012)
2. Bacaër, N., Guernaoui, S.: The epidemic threshold of vector-borne diseases with seasonality. J. Math. Biol. **53**, 421–436 (2006)
3. Bai, Z., Zhou, Y., Zhang, T.: Existence of multiple periodic solutions for an SIR model with seasonality. Nonlinear Anal. **74**, 3548–3555 (2011)
4. Boatto, S., Bonnet, C., Cazelles, B., Mazenc, F.: SIR model with time dependent infectivity parameter: approximating the epidemic attractor and the importance of the initial phase. HAL preprint, pp. 1–30 (2018)
5. Bohl, P.: Über differentialgleichungen. J. für Reine und Angewandte Mathematik **144**, 284–313 (1913)
6. Chueshov, I.: Monotone Random Systems Theory and Applications. LNM, vol. 1779. Springer, Heidelberg (2002). https://doi.org/10.1007/b83277
7. Daleckii, J., Krein, M.: Stability of Solutions of Differential Equations in Banach Space. Translations of Mathematical Monographs, vol. 43. AMS, Providence (1974)
8. Evans, L.C.: An Introduction to Stochastic Differential Equations, vol. 82. American Mathematical Society, Providence (2012)
9. Gao, S., Kong, S., Chen, W., Clarke, E.: Delta-complete analysis for bounded reachability of hybrid systems. CoRR abs/1404.7171 (2014)
10. Henzinger, T.: The theory of hybrid automata. In: Inan, M.K., Kurshan, R.P. (eds.) Verification of Digital and Hybrid Systems. NATO ASISeries F: Computer and Systems Sciences, vol. 170, pp. 265–292. Springer, Heidelberg (2000). https://doi.org/10.1007/978-3-642-59615-5_13
11. Kermack, W., McKendric, A.: A contribution to the mathematical theory of epidemics. Proc. Royal Soc. Lond. A Math. Phys. Eng. Sci. **772**(115), 700–721 (1927)
12. Kloeden, P.E., Pötzsche, C.: Nonautonomous bifurcation scenarios in SIR models. Math. Methods Appl. Sci. **38**(16), 3495–3518 (2015). https://doi.org/10.1002/mma.3433
13. Kloeden, P., Kozyakin, V.: The dynamics of epidemiological systems with nonautonomous and random coefficients. MESA **2**(2), 159–172 (2011)
14. Kong, S., Gao, S., Chen, W., Clarke, E.: dReach: δ-reachability analysis for hybrid systems. In: Baier, C., Tinelli, C. (eds.) TACAS 2015. LNCS, vol. 9035, pp. 200–205. Springer, Heidelberg (2015). https://doi.org/10.1007/978-3-662-46681-0_15
15. Kong, S., Solar-Lezama, A., Gao, S.: Delta-decision procedures for exists-forall problems over the reals. In: Chockler, H., Weissenbacher, G. (eds.) CAV 2018. LNCS, vol. 10982, pp. 219–235. Springer, Cham (2018). https://doi.org/10.1007/978-3-319-96142-2_15

16. Kuniya, T.: Existence of a nontrivial periodic solution in an age-structured SIR epidemic model with time periodic coefficients. Appl. Math. Lett. **27**, 15–20 (2014)
17. Lyapunov, A.: The General Problem of the Stability of Motion. Ann. Math. Studies, vol. 17. Taylor and Francis, Princeton (1949). English reprint (edn.)
18. Ponciano, J., Capistán, M.: First principles modeling of nonlinear incidence rates in seasonal epidemics. PLOS Comput. Biol. **7**(2), E1001079 (2011)
19. Potzsche, C.: Nonautonomous bifurcation of bounded solutions ii: a shovel bifurcation pattern. Discrete Contin. Dyn. Syst. (Ser. A) **31**(1), 941–973 (2010)
20. Raskin, J.: An introduction to hybrid automata. In: Hristu-Varsakelis, D., Levine, W.S. (eds.) Handbook of Networked and Embedded Control Systems. Control Engineering. Birkhäuser, Boston (2005)
21. Rocha, E., Silva, C., Torres, D.: The effect of immigrant communities coming from higher incidence tuberculosis regions to a host country. Ricerche di Matematica **67**(1), 89–112 (2018)
22. Sacker, R., Sell, G.: A spectral theory for linear differential systems. J. Differ. Eqn. **27**, 320–358 (1978)
23. Wang, W., Zhao, X.: Threshold dynamics for compartmental epidemic models in periodic environments. J. Dyn. Diff. Eqn. **20**, 699–717 (2008)
24. Wu, H., Li, W.: Poincaré type theorems for non-autonomous systems. J. Differ. Eqn. **245**, 2958–2978 (2008)

Combinatorial Dynamics
for Regulatory Networks

Zane Huttinga, Bree Cummins, and Tomas Geadon[(⊠)]

Department of Mathematical Sciences, Montana State University,
Bozeman, MT 59715, USA
tgedeon@montana.edu

Abstract. Modeling dynamics of cellular networks presents significant
challenges due to ill-defined nonlinearities, poorly characterized parameters, and noisy experimental data. Switching ODE systems are a modeling platform based on Boolean networks that allow the combinatorialization of phase space and parameter space, which in turn allows
the computation of summaries of global dynamics across all parameters.
In this contribution, we expand the class of cellular processes to those
with regulated degradation (RD systems). We show that RD systems
also admit a finite combinatorialization of phase space. We then show
that a special RD system, called a mixed PTM-switching system, admits
the combinatorialization of the parameter space that is represented as
a parameter graph. This graph is directly comparable to the parameter
graph of the switching system model of the same regulatory network. By
comparing dynamics that correspond to the same parameter graph node,
we show that the mixed system admits only a subset of the dynamics of
the switching system. Finally, we address the relationship between combinatorial representatives of dynamics and trajectories of the underlying
ODE systems. We provide necessary and sufficient conditions that guarantee that representatives of equilibria correspond to true equilibria of
the dynamics.

1   Introduction

Over the last 20 years the ability to sequence genomes has raised the question
of how these genes orchestrate highly efficient cellular responses to external signals. The cellular actors like genes, proteins, and signaling molecules can be
organized into networks by recording their directed interactions. While the type
of interaction between two such actors (positive or negative interaction) can be
experimentally established relatively easily, it is difficult to measure the rates of
these interactions. This presents significant challenges for modeling the dynamics of gene regulatory networks, since the choice of type of nonlinear model is
not governed by any first principle physical law. Nonlinearities are chosen either
assuming mass action kinetics, or assuming enzymatic reactions that lead to Hill
functions. Since initial conditions are usually not known precisely and parameters are only known to some loose bound, the interrogation of the model often

© Springer Nature Switzerland AG 2019
M. Chaves and M. A. Martins (Eds.): MLCSB 2018, LNCS 11415, pp. 56–73, 2019.
https://doi.org/10.1007/978-3-030-19432-1_4

consists of sampling as many parameters and initial conditions as feasible, running the model for a finite time, and comparing the results to the data. The comparison between uncertain, imprecise data and a sparsely-sampled, precise model presents significant challenges.

To address these difficulties, several more qualitative approaches have been proposed. Boolean networks [1,2,16,18,20,21] represent the state of each gene as either on or off. The update rule in these models could be either synchronous, where all genes update their state at the same time based on their current inputs, or asynchronous [3,22]. These models have a reduced number of parameters and are conceptually simple, but it is challenging to falsify them by data. A related concept of *switching* ODE systems [4,9–12,14,15] were introduced in the 1970s as a way to insert continuous time dynamics into Boolean networks by setting the right hand side of the ODE to a Boolean function, with the addition of a decay term. The advantage of this approach is the combinatorialization of the phase space: the behavior of the solutions only depends on where the solution is with respect to all thresholds of all nonlinearities. The number of states is finite, so the enumeration of allowable sequences of states is computable. The extent to which such a sequence of states corresponds to a solution of a system of ODEs has been resolved in some cases, but is, in our mind, secondary. We view a set of potential sequences of states as a valid description of a system trajectory. This is summarized by a *state transition graph* that records all possible nearest-neighbor transitions and encapsulates a description of global dynamics.

This view naturally leads to a characterization of dynamics in terms of the strongly connected components of the state transition graph, as well as reachability between them. These are encoded as the *Morse nodes and edges* of a *Morse graph*, respectively. Furthermore, the parameter space of a switching system decomposes into a finite number of parameter regions, where the Morse graph is identical for all parameters within that region. These regions can be computed analytically a priori, without any simulation of the system, and they are encoded as nodes of a *parameter graph*. The edges of the parameter graph represent the geometrical proximity of the corresponding pair of regions separated by a codimension-1 hyperplane [6–8,13].

Given the success of the combinatorial approach, it is natural to ask to what extend it is dependent on the particular form of a switching system, and if it is possible to extend it to other types of models. In our previous work we have shown that models that contain chains of linear equations [17] are also amenable to this approach. This is particularly important for applications where both mRNA and protein are modeled, because protein growth rate depends linearly on mRNA concentration. We have also considered the extension of the methodology to nonlinearities that are ramp-like and, unlike switching functions, have a continuous function connecting two constant values [5]. In both cases it is possible to define state transition graphs, Morse graphs and finite parameter graphs.

In this work we consider another important extension that admits a state transition graph and a parameter graph. Here we add equations that describe a modification of one protein by another. This encompasses both the case when the decay rate is regulated by another protein, such as ubiquitination, and other post-transcriptional modifications like phosphorylation.

Our first and most important result is that these systems, which we call *switching systems with regulated degradation*, or *RD systems*, admit the combinatorialization of both phase space and the parameter space. While combinatorialization of the phase space has been recognized previously [19], this is the first time where the state transition graph is explicitly constructed and it is shown that parameter space can be decomposed to a finite number of parameter regions, each of which has the same state transition graph.

We then proceed to analyze differences in dynamics that regulated degradation may bring. In particular, we consider the extreme case of the RD system that arises from modeling some types of post-transcriptional modification (PTM). This leads to a Lotka-Volterra type equation, where the growth term is switch-like. We first compare the dynamics of a switching system and a system where some switching equations are replaced by PTM equations. We show on an example that the PTM system has a restricted set of potential dynamics in comparison to the switching system. In switching systems, a particular type of Morse node, which will be denoted by FP, implies the existence of a stable equilibrium in the corresponding domain of the phase space. Since the PTM equations, unlike switching systems, may lead to unbounded dynamics, we prove necessary and sufficient conditions for FP Morse nodes in PTM system to contain stable equilibria.

2   Regulatory Networks and RD Switching Systems

Definition 2.1. A *regulatory network* is a finite, directed, annotated graph (V, E) with vertex set $V = \{1, \ldots, N\}$ and edge set $E \subset V \times V \times \{\rightarrow, \dashv\}$. An edge $(i, j, \rightarrow) \in E$ indicates that i positively regulates (activates) j, while $(i, j, \dashv) \in E$ indicates that i negatively regulates (represses) j. We assume that for each $i, j \in V$, there is at most one edge from i to j in E and we will use a shorthand $i \rightarrow j$ or $i \dashv j$ to denote positive (negative) regulation, respectively. We will assume that a regulatory network does not include any negative self-regulation, so for each $i \in V$, $(i, i, \dashv) \notin E$. We allow positive self-regulation, $i \rightarrow i$. For each $i \in V$, we define the *sources* and *targets* of i to be the sets $S(i) := \{s \in V \mid (s, i) \in E\}$ and $T(i) := \{j \in V \mid (i, j) \in E\}$, respectively.

Definition 2.2. An *RD switching* system on a regulatory network (V, E) is a system of ordinary differential equations

$$\dot{x}_j = \Lambda_j(x) - x_j \Gamma_j(x), \quad j = 1, \ldots, N \tag{1}$$

where $x = (x_1, \ldots, x_N)$ and for each $j \in V$, the functions Γ_j and Λ_j are defined in the following way. For $j \in V$ and each $i \in S(j)$, define a map $\sigma_{j,i} : \mathbb{R} \rightarrow \{l_{j,i}, u_{j,i}\}$ as

$$\text{If } i \to j, \ \sigma_{j,i}^{+}(x) = \begin{cases} l_{j,i} & \text{if } x_i < \theta_{j,i} \\ u_{j,i} & \text{if } x_i > \theta_{j,i} \\ \text{undefined} & \text{otherwise} \end{cases}, \text{ and if } i \dashv j, \ \sigma_{j,i}^{-}(x) = \begin{cases} u_{j,i} & \text{if } x_i < \theta_{j,i} \\ l_{j,i} & \text{if } x_i > \theta_{j,i} \\ \text{undefined} & \text{otherwise} \end{cases},$$

where $0 < l_{j,i} < u_{j,i}$ are constants and $\theta_{j,i}$ is a *threshold value* associated to the edge (i,j). We will use notation σ when denoting either of the functions $\sigma^{\pm}$.

Consider a decomposition of the set of sources of a vertex j $S(j;\Gamma) \cup S(j;\Lambda) = S(j)$. Define $\sigma_j^{\Gamma} : \mathbb{R}^{|V|} \to \mathbb{R}^{|S(j;\Gamma)|}$ and $\sigma_j^{\Lambda} : \mathbb{R}^{|V|} \to \mathbb{R}^{|S(j;\Lambda)|}$ component-wise by

$$\sigma_j^{\Gamma}(x) = (\sigma_{j,i}(x))_{i \in S(j;\Gamma)} \text{ and } \sigma_j^{\Lambda}(x) = (\sigma_{j,i}(x))_{i \in S(j;\Lambda)}.$$

At each node j we specify the interaction of the input variables in $S(j;\Gamma)$ that form the value of the function Γ, at the same time we specify the interaction of the input variables in $S(j;\Lambda)$ that form the value of the function Λ. These interactions are specified by *logic functions* $M_j^{\Gamma} : \mathbb{R}^{|S(j;\Gamma)|} \to \mathbb{R}$ and $M_j^{\Lambda} : \mathbb{R}^{|S(j;\Lambda)|} \to \mathbb{R}$. We assume that these functions are multilinear and thus take the form

$$M_j(\xi_1, \ldots, \xi_t) = \prod_{k=1}^{s} \sum_{i \in P_k^j} \xi_i \text{ where } P_k^j \subset S(j) \text{ and } P_a^j \cap P_b^j = \emptyset \text{ for } a \neq b.$$

For the function Γ, we set

$$\Delta_j := M_j^{\Gamma} \circ \sigma_j^{\Gamma} \quad \text{and} \quad \Gamma_j := \pm \Delta_j + \gamma_j, \tag{2}$$

where γ_j is a constant decay rate associated to the node j. For the function Λ we set

$$\Lambda_j := M_j^{\Lambda} \circ \sigma_j^{\Lambda}. \tag{3}$$

There are several biological situations that fit within our framework.

1. First, when $\Gamma_j := \Delta_j + \gamma_j$, the differential equation is

$$\dot{x}_j = \Lambda(x) - \Delta_j(x)x_j - \gamma_j x_j$$

   and the term $-\Delta_j(x)x_j$ represents a targeted degradation term.
2. Second, consider a molecular species x with total abundance x_{tot} and active (x_{act}) and inactive (x_{inac}) forms, with $x_{act} + x_{inac} = x_{tot}$. Assume that x_{inac} is activated by a PTM mechanism (say phosphorylation) by another molecule y and that this process is modeled by a switching function $\Upsilon(y)$. We also assume non-specific decay rate γ. Then

$$\begin{aligned} \dot{x}_{act} &= \Upsilon(y)x_{inac} - \gamma x_{act} = \Upsilon(y)(x_{tot} - x_{act}) - \gamma x_{act} \\ &= \Upsilon(y)x_{tot} - (\Upsilon(y) + \gamma)x_{act} \\ &=: \Lambda(y) - \Gamma(y)x_{act}, \end{aligned}$$

   where $\Gamma(y) := \Upsilon(y) + \gamma$ and $\Lambda(y) := \Upsilon(y)x_{tot}$.

3. Now consider PTM modification process close to a chemical equilibrium where the total of active and inactive forms are conserved $x_{act} + x_{inac} = x_{tot}$, and there is no non-specific decay rate γ. The basic equation for abundance of x_{act} reads $\dot{x}_{act} = \Upsilon(y)x_{inac} - kx_{act}$ and the equilibrium is given by $kx^*_{act} = \Upsilon(y)x^*_{inac}$. Introducing new variables $a := x_{act} - x^*_{act}, b := x_{inac} - x^*_{inac}$ the conservation equation reads

$$x_{tot} = x_{inac} + x_{act} = x^*_{inac} + b + x^*_{act} + a$$

which implies $b = -a$. The differential equation then reads

$$\dot{a} = \Upsilon(y)(x^*_{inac}+b)-k(x^*_{act}+a) = \Upsilon(y)b-ka = -\Upsilon(y)a-ka =: -a(-\Delta(y)+k) =: -\Gamma(y)a.$$

4. Finally, consider a particular limiting case of the PTM process, where the y-mediated activation is dominated by de-activation with rate k, $\Upsilon(y) <<$ k with non-specific decay rate $\gamma > 0$ Then, close to chemical equilibrium $kx_{act} \approx \Upsilon(y)x_{inac}$. It follows that

$$x_{tot} = x_{inac} + x_{act} \approx x_{inac} \quad \text{and} \quad x_{tot} = x_{inac} + x_{act} = (\frac{\Upsilon(y) + k}{\Upsilon(y)})x_{act},$$

and thus

$$\dot{x}_{act} = \Upsilon(y)x_{inac} - \gamma x_{act} \approx \Upsilon(y)x_{tot} - \gamma x_{act} \approx \Upsilon(y)(\frac{\Upsilon(y) + k}{\Upsilon(y)}x_{act}) - \gamma x_{act}$$

$$= (\Upsilon(y) + k)x_{act} - \gamma x_{act} =: -x_{act}(-\Delta(y) + \gamma) =: -\Gamma(y)x_{act}$$

We allow the sets $S(j;\Gamma)$ or $S(j;\Lambda)$ to be empty. This gives rise to two special cases of (1).

1. If for some $j \in V$, $S(j,\Gamma) = \emptyset$, then we call j a *switching node* and x_j a *switching variable*. In this case, $M^\Gamma_j : \{0\} \hookrightarrow \mathbb{R}$, so that $\Delta_j(x) = 0$ and (1) becomes

$$\dot{x}_j = \Lambda_j(x) - \gamma_j x_j.$$

If every variable is a switching variable, then we have the well-known *switching system*.

2. If for some $j \in V$ $S(j;\Lambda) = \emptyset$ and $\Gamma_j = -\Delta_j + \gamma_j$, then the (1) becomes

$$\dot{x}_j = -x_j(-\Delta_j(x) + \gamma_j).$$

This type of equation models post-transcriptional modification of x_j under assumption (3) or (4) above. We will call j a *PTM node* and x_j a *PTM variable*. A system in which every variable is a PTM variable is called a *PTM system*. One in which each variable is either switching or PTM is called a *PTM-switching* system.

For any regulatory network (V, E) Definition 2.2 assigns to each node $j \in V$ a parameter $\gamma_j > 0$, the decay rate, and to each edge $(i, j) \in E$ three positive parameters: the threshold $\theta_{j,i}$ and the two constants $l_{j,i}$ and $u_{j,i}$, where $l_{j,i} < u_{j,i}$ denote *lower* and *upper* values, respectively.

Definition 2.3. Consider a regulatory network (V, E) and its associated system (1). A *parameter* for the system is a tuple $z := (l, u, \theta, \gamma)$ of length $|V| + 3|E|$, where

- $l := \{l_{j,i} \mid (i,j) \in E\}$, $u := \{u_{j,i} \mid (i,j) \in E\}$, $\theta := \{\theta_{j,i} \mid (i,j) \in E\}$, and $\gamma := \{\gamma_j \mid j \in V\}$;
- $\gamma_i > 0$ for each node $i \in V$, and $0 < l_{j,i} < u_{j,i}$ and $0 < \theta_{j,i}$ for every edge $(i,j) \in E$; and
- for each $i \in V$, the thresholds $\theta_{j,i}$, $j \in T(i)$ are distinct.

With the last assumption, which is generic, we avoid unnecessary degeneracies in both the construction of the state transition graph and the parameter graph. Since our description of dynamics is based on relative inequalities between values $u_{j,i}, l_{j,i}$ and thresholds $\theta_{k,j}$ and not the absolute value of their differences, any difference between the thresholds is sufficient. Note that this assumption allows both $\Lambda(x)$ and $\Gamma(x)$ to change at the same threshold, if for an edge $(i,j) \in E$, we have $i \in S(j; \Gamma) \cap S(j; \Lambda)$.

The thresholds $\{\theta_{j,i}\}$ for a node $i \in V$ form a total order $\theta_{j_1,i} < \cdots < \theta_{j_{T(i)},i}$, and we define the order $O_i(z) := j_1 < j_2 < \cdots < j_{T(i)}$. The collection of all orders is denoted $O(z) := \{O_i(z) \mid i \in V\}$.

A collection of all thresholds at a parameter z decomposes the phase space of (1) into a finite set of non-empty domains.

Definition 2.4. Consider regulatory network (V, E) and its associated system (1). Fix a parameter $z := (l, u, \theta, \gamma) \in Z$. We will adopt the notational convention that for each $i \in V$, $\theta_{-\infty,i} := 0$ and $\theta_{\infty,i} := \infty$. We say two thresholds $\theta_{j,i}$ and $\theta_{k,i}$ are *consecutive* if $\theta_{j,i} < \theta_{k,i}$ and there is no $l \in T(i)$ such that $\theta_{j,i} < \theta_{l,i} < \theta_{k,i}$.

For each i, assume $\theta_{v_i,i} < \theta_{w_i,i}$ are consecutive thresholds. Then a *domain* is a subset of the phase space $(0, \infty)^N$,

$$\kappa := \prod_{i=1}^{N} (\theta_{v_i,i}, \theta_{w_i,i}).$$

We denote the set of all domains in phase space at the parameter z by $\mathcal{K}(z)$. A *face* of $\kappa \in \mathcal{K}(z)$ is a set

$$\tau := \prod_{i=1}^{j-1} (\theta_{v_i,i}, \theta_{w_i,i}) \times \{\theta_{k,j}\} \times \prod_{i=j+1}^{N} (\theta_{v_i,i}, \theta_{w_i,i}).$$

where $k \in T(j)$ and so the threshold $\theta_{k,j}$ is nonzero and finite. We say τ is a *left face* of κ if $k = v_j$ and a *right face* if $k = w_j$.

Because each domain is bounded by two consecutive thresholds in each direction, for each $i \in V$, Λ_i and Γ_i are constant on κ. We denote these constants $\Lambda_j(\kappa)$ and $\Gamma_j(\kappa)$, respectively. The ODE system reduces to a set of decoupled affine equations

$$\dot{x}_j = \Lambda_j(\kappa) - x_j \Gamma_j(\kappa).$$

When $\Gamma_j(\kappa) > 0$, the analytic solution to the affine equation shows $\lim_{t\to\infty} x_j(t) = \Lambda_j(\kappa)/\Gamma_j(\kappa)$, as long as $x_j(t) \in \kappa$ for all $t \geq 0$. When $\Gamma_j(\kappa) < 0$, then $x_j(t) \to \infty$ as $t \to \infty$.

Definition 2.5. We say that

$$\Lambda(\kappa)/\Gamma(\kappa) := (\lambda_1, \ldots, \lambda_N)$$

is the *focal point of* κ, where $\lambda_i = \Lambda_i(\kappa)/\Gamma_i(\kappa)$ when $\Gamma_i(\kappa) > 0$, and $\lambda_i = \infty$ otherwise. If $\Lambda(\kappa)/\Gamma(\kappa) \in \kappa$, then we say that κ is an *attracting domain*. Clearly, any solution trajectory beginning in κ must remain in κ for all time.

Definition 2.6. Let (V, E) be a regulatory network with an associated system (1). We say a parameter z is *regular* if for each $i \in V$, for each $j \in T(i)$, $\Lambda_i(\kappa) - \theta_{j,i}\Gamma_i(\kappa) \neq 0$. We denote the set of all regular parameters by Z.

Definition 2.7. Let (V, E) be a regulatory network and fix a regular parameter $z \in Z$. A *wall* is a pair (τ, κ) where $\kappa \in \mathcal{K}(z)$ is a domain and τ is a face of κ. The set of all walls at the parameter z is denoted $\mathcal{W}(z)$.

For each wall $(\tau, \kappa) \in \mathcal{W}(z)$, we define the *sign* of the wall to be

$$\text{sgn}\,(\tau, \kappa) = \begin{cases} -1 & \text{if } \tau \text{ is a right face of } \kappa \\ 1 & \text{if } \tau \text{ is a left face of } \kappa \end{cases}.$$

We define the *wall-labeling* function $\ell : \mathcal{W}(z) \to \{-1, 1\}$ where for each $(\tau, \kappa) \in \mathcal{W}(z)$, if $\tau \subset \{x_i = \theta_{j,i}\}$, then

$$\ell(\tau, \kappa) = \text{sgn}\,(\tau, \kappa) \cdot \text{sgn}\,(\Lambda_i(\kappa) - \theta_{j,i}\Gamma_i(\kappa)). \tag{4}$$

We say (τ, κ) is an *incoming wall* if $\ell(\tau, \kappa) = 1$ and an *outgoing wall* if $\ell(\tau, \kappa) = -1$.

We illustrate the wall labeling function on an example that we will use throughout the paper. We initially endow the network in Fig. 1 a with a switching system, and will later endow it with a PTM-switching system.

Example 2.8. Consider the regulatory network (V, E) with $V = \{1, 2\}$ and edges $1 \to 1$, $1 \dashv 2$, and $2 \to 1$ (Fig. 1a). We endow the network with with a switching system $\dot{x}_i = -\gamma_i x_i + \Lambda_i(x)$, $i = 1, 2$, where $M_1(a, b) = a + b$:

$$\Lambda_1(x) = \begin{cases} l_{1,1} + l_{1,2} \text{ if } x_1 < \theta_{1,1} \text{ and } x_2 < \theta_{1,2} \\ l_{1,1} + u_{1,2} \text{ if } x_1 < \theta_{1,1} \text{ and } x_2 > \theta_{1,2} \\ u_{1,1} + l_{1,2} \text{ if } x_1 > \theta_{1,1} \text{ and } x_2 < \theta_{1,2} \\ u_{1,1} + u_{1,2} \text{ if } x_1 > \theta_{1,1} \text{ and } x_2 > \theta_{1,2} \end{cases} \qquad \Lambda_2(x) = \begin{cases} u_{2,1} \text{ if } x_1 < \theta_{2,1} \\ l_{2,1} \text{ if } x_1 > \theta_{2,1} \end{cases}$$

and $\gamma_1 = \gamma_2 = 1$. In order to specify wall labeling in all domains κ, we need to specify inequalities between the values of the function Λ_1, evaluated on the inputs to node 1, and the products $\gamma_1\theta_{1,1}, \gamma_1\theta_{2,1}$ that correspond to the output edges of node 1. At the same time, we need to specify inequalities between the values of the function Λ_2, evaluated on the input to node 2, and the product $\gamma_2 \theta_{1,2}$ that corresponds to the output edge of node 2. We select the following inequalities which define a parameter region in the parameter space.

$$l_{1,1} + l_{1,2} < l_{1,1} + u_{1,2} < \gamma_1 \theta_{1,1} < u_{1,1} + l_{1,2} < \gamma_1 \theta_{2,1} < u_{1,1} + u_{1,2} \quad (5)$$
$$l_{2,1} < \gamma_2 \theta_{1,2} < u_{2,1}, \qquad \theta_{1,1} < \theta_{2,1}$$

For all parameters z in this region the wall labeling is the same and depicted in Fig. 1b. Since a wall consists of a pair (domain, face), the single-tipped arrows in Fig. 1b indicate that a face is an incoming wall with respect to the domain containing the tip and an outgoing wall with respect to the domain containing the tail. The double-tipped arrows indicate that the face is an incoming wall with respect to both neighboring domains.

(a) Regulatory network.

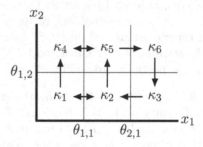

(b) Phase space with wall labeling.

Fig. 1. Regulatory network and wall labeling for the parameter values (5) in Example 2.8.

3   State Transition Graph and Its Morse Graph

We will now obtain a finite description of the coarse dynamics of (1). We begin by constructing a state transition graph. By extracting the strongly connected components of this state transition graph we build a condensed representation of the dynamics in the form of a Morse graph. We begin with two equivalent characterizations of the state transition graph, as a directed graph and as a multivalued map. For this section, we will fix a regulatory network (V, E) and an associated RD system.

Definition 3.1. Let $z \in Z$ be a regular parameter. Define the vertex set $\mathcal{V}$ of the state transition graph $(\mathcal{V}, \mathcal{E})$ as the image of the set of domains $\mathcal{K}(z)$ via a bijection $f : \mathcal{K}(z) \to \mathcal{V}$. We define a multivalued map $\mathcal{F} : \mathcal{V} \rightrightarrows \mathcal{V}$ that captures the dynamics of (1) as follows:

1. if $u = v$, then $v \in \mathcal{F}(u)$ if and only if $f^{-1}(u)$ is an attracting domain;
2. if $u \neq v$, then $v \in \mathcal{F}(u)$ if and only if there exists a face τ such that $(\tau, f^{-1}(u))$ is an outgoing wall and $(\tau, f^{-1}(v))$ is an incoming wall.

Given $\mathcal{F}$, for any pair of vertices $u, v \in \mathcal{V}$ there is a directed edge $(u, v) \in \mathcal{E}$, if and only if $v \in \mathcal{F}(u)$.

We now define a *Morse graph* that captures the essential parts of the dynamics of the state transition graph.

Definition 3.2. A strongly connected path component of a directed graph $\mathcal{G}$ is a maximal subgraph $\mathcal{C}$ of $\mathcal{G}$ such that for any vertices $u, v \in \mathcal{C}$, there exists a nonempty path in $\mathcal{C}$ from u to v. We will also refer to a strongly connected path component as a *recurrent component*. Let $\{\mathcal{C}_s\}_{s \in \mathsf{P}}$ be a collection of recurrent components, indexed by a set P. Define a partial order $\mathcal{P}$ on the set of Morse nodes $\{\mathcal{C}_s\}$ via the reachability relation on $(\mathcal{V}, \mathcal{E})$. In $\mathcal{P}$, we set $q \leq_{\mathcal{P}} p$ if and only if there exists a path in $(\mathcal{V}, \mathcal{E})$ from an element of $\mathcal{C}_p$ to an element of $\mathcal{C}_q$. The *Morse graph* of $(\mathcal{V}, \mathcal{E})$, denoted $\mathsf{MG}(\mathcal{V}, \mathcal{E})$, is the Hasse diagram of the vertex set $\{\mathcal{C}_s\}_{s \in \mathsf{P}}$ partially ordered by $\mathcal{P}$.

Remark 3.3. Note that if κ is an attracting domain, then $f(\kappa) \in \mathcal{V}$ has a self-edge by Definition 3.1. By Definitions 2.7 and 3.1, there can be no other out-going edges from κ, and so $f(\kappa)$ is a minimal Morse node of the Morse graph in the sense that it has no outgoing paths to any other Morse nodes.

As remarked earlier, all solutions of the ODE system remain in an attracting domain κ for all $t \geq 0$. In a switching system components of the target point are finite $f(\kappa)_i < \infty$ for all $i = 1, \ldots, N$ and thus the focal point of κ represents a unique equilibrium of (1). We will discuss the correspondence between attracting domains and equilibria in PTM-switching systems in Sect. 5.

We adopt a labeling system for the Morse nodes of the Morse graph $\mathsf{MG}(\mathcal{V}, \mathcal{E})$ that distinguishes Morse nodes that are composed of a single vertex in the state transition graph from those that have multiple vertices.

Definition 3.4. Let $\mathcal{C}$ be a node in $\mathsf{MG}(\mathcal{V}, \mathcal{E})$, that is, $\mathcal{C}$ is a recurrent component of the domain graph $(\mathcal{V}, \mathcal{E})$. Then if $\mathcal{C}$ consists of a single node of the state transition graph, we give $\mathcal{C}$ the label FP, for "fixed point vertex". We will call such Morse nodes FP-nodes. If $\mathcal{C}$ is not an FP, then we will label it C for "cycle", since recurrent paths through multiple nodes indicate potential recurrent trajectories.

Example 2.8 **Continued.** The state transition graph for the switching system is depicted in Fig. 2a. This gives rise to the Morse graph in Fig. 2b. There are two Morse nodes, one consisting of node ξ_4 denoted by FP and one cycle node, corresponding to nodes $\xi_3, \xi_2, \xi_5, \xi_6$ denoted by C (see Definition 3.4).

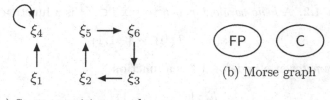

(a) State transition graph.

(b) Morse graph

Fig. 2. (a) state transition graph and (b) Morse graph for the switching system in Example 2.8 at the parameter node of the parameter graph given by (5).

4   Combinatorial Parameters and Parameter Graph

While the definitions of the state transition graph and Morse graph in the previous sections are valid for arbitrary RD systems, in this section we concentrate on the special case of PTM-switching systems. We consider a regulatory network (V, E) together with vertex sets $V^s \cup V^p = V$, $V^s \cap V^p = \emptyset$, that denote switching and PTM nodes, respectively.

We show how to assign to each parameter $z \in Z$ a combinatorial description $\phi := \omega(z)$ which is sufficient to construct the wall-labeling (and hence state transition diagram and Morse graph) induced by z. We call this combinatorial description a *combinatorial parameter* and we denote the collection of combinatorial parameters by Φ. The collection Φ induces a decomposition of parameter space into a finite number of regions, where every parameter within one of these regions has the same Morse graph, and hence the same dynamical description. We represent Φ as a *parameter graph* where combinatorial parameters ϕ and ϕ' are joined by an edge, if the corresponding regions in the parameter space share a co-dimension 1 boundary.

Definition 4.1. Define the *input combinations* of the node x_i to be the Cartesian product

$$\mathsf{In}_j := \prod_{i \in \mathsf{S}(j)} \{0, 1\}.$$

Define the *indicator function* $\chi_j : (0, \infty)^N \to \mathsf{In}_j$ such that

$$\chi_{j,i}(x) = \begin{cases} 0 & \text{if } i \to j \text{ and } x_i < \theta_{j,i} \text{ or if } i \dashv j \text{ and } x_i > \theta_{j,i} \\ 1 & \text{if } i \to j \text{ and } x_i > \theta_{j,i} \text{ or if } i \dashv j \text{ and } x_i < \theta_{j,i} \\ \text{undefined} & \text{otherwise.} \end{cases}$$

Define the *valuation function* $v_j : \mathsf{In}_j \to \mathbb{R}^{Sj}$ via

$$v_{j,i}(A) = \begin{cases} l_{j,i} & \text{whenever } A_i = 0 \\ u_{j,i} & \text{whenever } A_i = 1 \\ \text{undefined} & \text{otherwise.} \end{cases}$$

Note that $\sigma_j = v_j \circ \chi_j$, where σ_j is defined in (2.2).

Definition 4.2. A *logic parameter* at a vertex $i \in V^s$ is a function

$$g_i : (\mathsf{In}_i \times T(i)) \to \{-1, 1\}.$$

A *logic parameter* at a node $i \in V^p$ is a function

$$h_i : \mathsf{In}_i \to \{-1, 1\}.$$

The collection $L = (L_1, \ldots, L_N)$ with $L_i := g_i$ for $i \in V^s$ and $L_i := h_i$ for $i \in V^p$ is the *logic parameter*. An *order parameter* O is a collection of total orderings O_i of $T(i)$ for each $i \in X$, see Definition 2.3. A *combinatorial parameter* is a pair $\phi = (L, O)$ where L is a logic parameter and O is an order parameter. We denote the collection of combinatorial parameters as Φ. The *combinatorial assignment function* $\omega : Z \to \Phi$ is given by $\omega(z) := (L, O)$ where $O = O(z)$ from Definition 2.3, and

$$L_i(A, B) = \mathrm{sgn}\left(\left(M_i^{\Lambda} \circ v_i(A) - \gamma_i \theta_{B,i}\right)\right) \text{ for all } i \in V^s; \tag{6}$$
$$L_i(A) = \mathrm{sgn}\left(\left(M_i^{\Gamma} \circ v_i(A) - \gamma_i\right)\right) \text{ for all } i \in V^p .$$

For all $z \in Z$, we say that $\omega(z)$ is *the combinatorial parameter associated to the parameter* z. The *parameter region associated with the combinatorial parameter* ϕ is given by $\omega^{-1}(\phi) \subset Z$. A combinatorial parameter $\phi \in \Phi$ is *realizable* if there exists $z \in Z$ such that $\phi = \omega(z)$.

Note that a parameter region $\omega^{-1}(\phi) \subset Z$ consists of all parameters that satisfy a set of inequalities relating the inputs to node k to the weighted thresholds corresponding to the outputs of node k, for every node $k \in V$. An example of such a region is (5).

Definition 4.3. Let $\phi = (L, O) \in \Phi$ be a realizable combinatorial parameter, and let $z \in \omega^{-1}(\phi) \subset Z$. We induce a wall-labeling on $\mathcal{W}(z)$ as follows. Let (τ, κ) be a wall with projection index i and switching index j. We say (τ, κ) is an *outgoing wall* with respect to ϕ if $L_i(\chi_i(\kappa), j) = -\mathrm{sgn}\left((\tau, \kappa)\right)$ and an *incoming wall* if $L_i(\chi_i(\kappa), j) = \mathrm{sgn}\left((\tau, \kappa)\right)$.

We show that the wall-labeling given in Definition 4.3 coincides with one given in Definition 2.7.

Theorem 4.4. *A realizable combinatorial parameter* $\phi \in \Phi$ *determines the wall labeling function* ℓ. *That is, for every domain* $\kappa = \prod_{i=1}^{N}(\theta_{v_i,i}, \theta_{w_i,i})$ *and its face* $\tau := \prod_{i=1}^{j-1}(\theta_{v_i,i}, \theta_{w_i,i}) \times \{\theta_{k,j}\} \times \prod_{i=j+1}^{N}(\theta_{v_i,i}, \theta_{w_i,i})$, *the value* $\ell(\kappa, \tau)$ *is constant over* $\omega^{-1}(\phi)$.

Proof. Let $z, z' \in Z$ and suppose $\omega(z) = \omega(z') = \phi$. We will show $\ell(\tau, \kappa)$ has the same value at both z and z'.

First note that because $\omega(z) = \omega(z')$, by definition of ω, $O(z) = O(z')$ there is an order preserving bijection α between thresholds at z to thresholds at z'. This induces a bijection $\alpha : \mathcal{K}(z) \to \mathcal{K}(z')$ such that if $\tau \subset \theta_{k,j}$ implies

$\alpha(\tau) \subset \alpha(\theta_{k,j})$. This implies that if $\kappa' = \alpha(\kappa)$, a particular threshold $\theta_{k,j}$ is either the upper bound of the jth component $(\theta_{v_j,j}, \theta_{w_j,j})$ or the lower bound for both κ and κ'. This means that a given face is either a left face or a right face for both κ and κ'. Therefore $\mathrm{sgn}(\kappa, \tau)$ has the same value at both κ and κ'.

Now note that whether j is a switching node or a PTM node does not depend on a choice of parameters; nor does the logic function, M_j^Λ or M_j^Γ. Suppose j is a switching node. Then it remains to show that

$$\mathrm{sgn}\left(\Lambda_j \left(\prod_{i=1}^N (\theta_{v_i,i}, \theta_{w_i,i})\right) - \gamma_j \theta_{k,j}\right)$$

has the same value at z and z'. Because $\omega(z) = \omega(z')$, for every $(A, B) \in \mathsf{In}_j \times T(j)$,

$$\mathrm{sgn}\left(M_j^\Lambda \circ v_j(A) - \gamma_j \theta_{B,j}\right)$$

is equal at z and z'. Hence the result follows because $k \in T(j)$ and $\Lambda_j = M_j^\Lambda \circ v_j \circ \chi_{j,i}$ by (3). If j is a PTM node, then we find that

$$\mathrm{sgn}\left(\Delta_j \left(\prod_{i=1}^N (\theta_{v_i,i}, \theta_{w_i,i})\right) - \gamma_j\right)$$

has the same value at z and z' because $\Delta_j = M_j^\Gamma \circ v_j$ by (2), and for every $A \in \mathsf{In}_j$, $\mathrm{sgn}\left(M_j^\Gamma \circ v_j(A) - \gamma_j\right)$ has the same value at z and z'.   □

Given $\phi \in \Phi$, Theorem 4.4 tells us that the state transition and Morse graphs are identical for all z such that $\omega(z) = \phi$, and so the study of a finite number of combinatorial parameters characterizes all possible dynamical behaviors of a switching or PTM-switching system over all parameter space.

5 Comparison of Fixed Point Dynamics

Given a regulatory network (V, E) we discuss the similarities and differences between the dynamics of a switching system and a PTM-switching system. More specifically, if we replace some switching nodes by PTM nodes, how does this affect the Morse graph? We start by formalizing the idea of this comparison.

Suppose two systems have the same regulatory network (V, E), but different decompositions of switching and PTM variables. In particular, assume that one system admits a decomposition of network nodes $V = V^s \cup V^p$ and the other system admits $V = W^s \cup W^p$, such that $W^s := V^s \setminus Q$, and $W^p := V^p \cup Q$ for some $Q \subset V^s$.

The regular parameter space is the same set Z in both systems. However the logic function L^W will be different from the original logic function L^V, because different components of L will be determined by the PTM versus switching equations. As a consequence, the set of combinatorial parameters Φ^W is different than the set of of combinatorial parameters Φ^V. However, there is an injective map $\Omega : \Phi^W \to \Phi^V$ given by

$$\Omega(L, O) = (\Omega(L), O), \qquad \Omega(L) = (\Omega(L_1), \ldots, \Omega(L_N)).$$

Consider a combinatorial parameter $\phi = (L, O) \in \Phi^W$. For all $i \in V \setminus Q$ we set $\Omega(L_i) = L_i$. For $i \in Q$, we have $L_i := h_i(\alpha), \alpha \in \mathsf{In}_i$, and we need to define $\Omega(L_i)$. We set

$$\Omega(L_i) = g_i(\alpha, j) := h_i(\alpha) \text{ for all} j \in \mathsf{Out}_i. \tag{7}$$

Lemma 5.1. *The map Ω is not surjective, unless for all $i \in Q$, i has a single outgoing edge.*

Proof. It follows from formula (7) that Ω is onto if, and only if, $j = 1$ for all $i \in Q$. The result follows. $\qquad\square$

Theorem 5.2. *If $\phi = (L, O) \in \Phi^W$ is realizable, then $\phi' = (L', O') := \Omega(\phi)$ is realizable and the wall-labeling $\ell_\phi = \ell_{\phi'}$ is the same at both ϕ and ϕ'. Consequently, the state transition graphs and Morse graphs are isomorphic at ϕ and ϕ'.*

Proof. To prove the realizability of ϕ', select $z \in \omega^{-1}(\phi)$, with $z = (l, u, \theta, \gamma)$. We define $z' \in Z$, $z' = (l', u', \theta', \gamma')$, as follows. Let $\gamma_i' := \gamma_i$ for all $i \in V$, let $l_{i,j}' := l_{i,j}$ and $u_{i,j}' := u_{i,j}$ for all $(i, j) \in E$ and for $i \in V \setminus Q$, and each $j \in T(i)$, let $\theta_{j,i}' := \theta_{j,i}$.

Now we consider $i \in Q$. Because z is a regular parameter, $\Delta_i(\kappa) \neq \gamma_i$ for all domains $\kappa \in \mathcal{K}(z)$. Moreover, because the range of Δ_i is a finite set, there is an interval $(\gamma_i - \varepsilon, \gamma_i + \varepsilon)$ that does not intersect the range of Δ_i. Let $\{\theta_{j,i}'\}$ be any set with the same cardinality as the set $\{\theta_{j,i}\}$, that satisfies the following: (a) $\{\theta_{j,i}' \mid j \in T(i)\} \subset (\gamma_i - \varepsilon, \gamma_i + \varepsilon)$, (b) each threshold is distinct, and (c) the ordering of $\{\theta_{j,i}'\}$ is the same as that of $\{\theta_{j,i}\}$.

This last choice implies that the collection of orders of all thresholds $O(z) = O(z')$ and by (a) and (b) z' is a regular parameter. Furthermore, for each $i \in V \setminus Q$, the logic of $\omega(z')$ at i is equal to L_i. Then, for $i \in Q$, we have that for each $(A, B) \in \mathsf{In}_i \times \mathsf{Out}_i$, by construction $\mathrm{sgn}\left(M_i^\Lambda \circ v_i(A) - \gamma_i \theta_{B,i}\right) = \mathrm{sgn}\left(M_i^\Gamma \circ v_i(A) - \gamma_i\right)$, when the left and right hand sides are evaluated at z' and z, respectively. Hence z' belongs to parameter region $\phi' = \omega(z')$ that satisfies $\phi' = \Omega(\phi)$.

By Theorem 4.4, ϕ and ϕ' both determine the wall-labeling function ℓ, which in turn determines the state transition and Morse graphs. Let (κ, τ) be a wall at parameter node ϕ, where $\kappa = \prod_{i=1}^N (\theta_{v_i,i}, \theta_{w_i,i})$ is a domain and $\tau := \prod_{i=1}^{j-1} (\theta_{v_i,i}, \theta_{w_i,i}) \times \{\theta_{k,j}\} \times \prod_{i=j+1}^N (\theta_{v_i,i}, \theta_{w_i,i})$ is a face of κ. Since $O = O'$ and the intervals $(\theta_{v_i,i}, \theta_{w_i,i})$ are all well-defined for parameters in parameter node ϕ for all $i \in V$, they are also well-defined at ϕ'. Similarly, if $j = v_k, (j = w_k)$ at ϕ, then the same is true at ϕ' as well. Therefore, if τ is a left (right) face of κ at ϕ, then it is a left (right) at ϕ'. Hence $\mathrm{sgn}(\tau, \kappa)$ is the same at ϕ and ϕ'.

Recall that if $i \in V \setminus Q$, then $L_i' = L_i$ by definition of Ω. Further, for each $i \in Q$, for each $(A, B) \in \mathsf{In}_i \times \mathsf{Out}_i$, $L_i'(A, B) = g_i(A, B) = h_i(A) = L_i(A)$. It follows that

$$\mathrm{sgn}\left(\Lambda_j(\kappa) - \gamma_j \theta_{k,j}\right) = \mathrm{sgn}\left(M_j^\Lambda \circ v_j(A) - \gamma_j \theta_{k,j}\right) = \mathrm{sgn}\left(M_j^\Gamma \circ v_j(A) - \gamma_j\right) = \mathrm{sgn}\left(\Delta_j(\kappa) - \gamma_j\right),$$

for an appropriate $A = \chi_{k,j}(\kappa)$. This together with the fact that $\mathrm{sgn}(\tau, \kappa)$ is the same at ϕ and ϕ' proves that the labeling function is the same at ϕ and ϕ'.   □

The map $\Omega : \Phi^W \to \Phi^V$ is an injective, but not necessarily surjective, map of combinatorial parameters by Lemma 5.1. This implies that the collection of combinatorial parameters of a switching system is richer than that of a PTM system. There is a natural question whether also the collection of types of Morse graphs $\mathsf{MG}_V(\mathcal{V}, \mathcal{E})$ exhibited in the parameter graph Φ^V, is richer than a collection of the Morse graphs $\mathsf{MG}_W(\mathcal{V}, \mathcal{E})$ of the parameter graph Φ^W

$$\mathsf{MG}_W(\mathcal{V}, \mathcal{E}) \subset \mathsf{MG}_V(\mathcal{V}, \mathcal{E}).$$

We now continue with Example (2.8) to show that the switching systems, indeed, have richer dynamics.

Example 2.8 Continued: In Fig. 2b we listed a Morse graph for a parameter node (5) in Φ^V. Consider the PTM-switching system obtained by making node 1 of the network in Fig. 1a into a PTM node:

$$\dot{x}_1 = x_1(\Delta_1(x) - \gamma_1)$$
$$\dot{x}_2 = -\gamma_2 x_2 + \Lambda_2(x).$$

We set $\Delta_1 := \Lambda_1$ from the original switching system of Example 2.8, and keep Λ_2 the same. By examining the set of all combinatorial parameters Φ^W, we observe that there exist only four Morse graphs that we show in Fig. 3(a)–(d). None of them are isomorphic to the switching system's Morse graph in Fig. 2b.

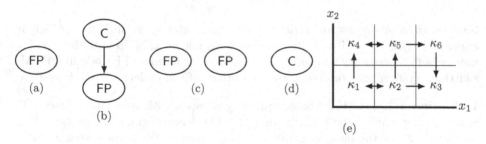

Fig. 3. (a)–(d): All Morse graphs for the PTM-switching system in Example 2.8. (e) Wall labeling for parameter (8).

As we have seen in Remark 3.3 each Morse node labeled FP corresponds to a unique stable equilibrium of a switching system. One of the principal differences between the equation describing the dynamics of a switching node and a PTM node is that the dynamics of the switching node are always bounded. This follows from the boundedness of Λ and the presence of the decay constant $\gamma > 0$.

As a consequence, a switching system always has a compact global attractor. On the other hand, PTM equations are not necessarily bounded when

$$\Delta(x) - \gamma > 0.$$

This presents challenges when one wants to deduce the existence of a recurrent set of the PTM-switching system from existence of a Morse node.

We illustrate this on our continuing example. Consider the PTM-switching system in Example 2.8 at parameter

$$l_{1,1} + l_{1,2} < l_{1,1} + u_{1,2} < \gamma_1 < u_{1,1} + l_{1,2} < u_{1,1} + u_{1,2} \tag{8}$$
$$l_{2,1} < \gamma_2 \theta_{1,2} < u_{2,1}, \qquad \theta_{1,1} < \theta_{2,1}$$

with two Morse nodes FP that correspond to domains κ_3 and κ_4 (see Fig. 3(e) for wall labeling and Fig. 3(c) for Morse graph). We will show next that there is no equilibrium in κ_3; in fact the trajectories in κ_3 all diverge to infinity.

To see this, note that on $\kappa_3 = (\theta_{2,1}, \infty) \times (0, \theta_{1,2})$, the system reduces to

$$\dot{x}_1 = x_1(u_{1,1} + l_{1,2} - \gamma_1), \qquad \dot{x}_2 = -\gamma_2 x_2 + l_{2,1},$$

and when $x(0) = (x_1(0), x_2(0)) \in \kappa_3$, this yields the following solution.

$$x_1(t) = x_1(0)e^{(u_{1,1}+l_{1,2}-\gamma_1)t}, \qquad x_2(t) = \left(x_2(0) - \frac{l_{2,1}}{\gamma_2}\right)e^{-\gamma_2 t} + \frac{l_{2,1}}{\gamma_2}.$$

Since in the combinatorial parameter (8) we have $u_{1,1} + l_{1,2} > \gamma_1$,

$$\lim_{t \to \infty} x_1(t) = \infty, \quad \lim_{t \to \infty} x_2(t) = \frac{l_{2,1}}{\gamma_2} \in (0, \theta_{1,2}).$$

Since both components are strictly monotonic, $x(t) \in \kappa_3$ for all t, which is consistent with the wall labeling, but κ_3 does not contain an equilibrium. We now present a necessary and sufficient condition for when an FP node in a PTM-switching system does represent an equilibrium of the underlying ODE system.

Definition 5.3. Let (V, E) be a regulatory network with associated system (1) where each variable is either switching or PTM. Fix a regular parameter $z \in Z$ and let $(\mathcal{V}, \mathcal{E})$ be the domain graph at this parameter. We say a vertex $\xi \in \mathcal{V}$ is a *degenerate* FP *node* if it is an FP Morse node and at least one component λ_i of the focal point of $\kappa := f^{-1}(\xi)$ is infinite, $\lambda_i = \infty$.

We now characterize degenerate nodes FP using only their location in phase space relative to the threshold hyperplanes $\{x_i = \theta_{j,i}\}$, $i \in V$, $j \in T(i)$.

Theorem 5.4. *An* FP *node ξ is degenerate if and only if there exists a PTM node $i \in V$ such that on $\kappa := f^{-1}(\xi)$, for each $j \in T(i)$ we have $\theta_{j,i} < x_i$.*

Proof. For the forward implication, assume a node ξ is degenerate. Then there is $i \in V$ such that i-th component of the target point $\lambda_i = \infty$ and therefore $\lim_{t \to \infty} x_i(t) = \infty$ whenever $x(0) \in \kappa = f^{-1}(\xi)$. Now suppose by contradiction that there is some $j \in T(i)$ such that on κ, $x_i < \theta_{j,i}$. Without loss of generality, assume that j is the target of i where $\theta_{j,i}$ is the smallest threshold for which this inequality holds. Let τ be the face of κ on which $x_i = \theta_{j,i}$. Then because κ is an attracting domain, (τ, κ) is an incoming wall with $\ell(\tau, \kappa) = 1$. Because τ is a right face of κ, $\text{sgn}(\tau, \kappa) = -1$. Hence $\text{sgn}(\theta_{j,i}(\Delta_i(\kappa) - \gamma_i)) = -1$; in particular, $\text{sgn}(\Delta_i(\kappa) - \gamma_i) = -1$.

Recall that when $x(0) \in \kappa$, the solution starting at $x(0)$ has the form

$$x_i(t) = x_i(0)e^{(\Delta_i(\kappa) - \gamma_i)t}.$$

Because we have shown that $\Delta_i(\kappa) - \gamma_i < 0$, the trajectory $x_i(t)$ approaches 0 as t approaches infinity. This contradicts our assumption that $\lim_{t \to \infty} x_i(t) = \infty$.

For the backward implication, let i be a PTM node such that for each $j \in T(i)$, $x_i > \theta_{j,i}$ on κ. Let $k \in T(i)$ satisfy $\theta_{j,i} \leq \theta_{k,i}$ for each $j \in T(i)$. Then if $\tau \subset \{x_i = \theta_{k,i}\}$, then τ is a left face of κ, meaning $\text{sgn}(\tau, \kappa) = 1$. Because κ is attracting, (τ, κ) is an incoming wall: $\ell(\tau, \kappa) = 1$. Therefore, $\text{sgn}(\theta_{k,i}(\Delta_i(\kappa) - \gamma_i)) = \text{sgn}((\Delta_i(\kappa) - \gamma_i)) = 1$. Therefore, using the above formula for $x_i(t)$, we have $\lim_{t \to \infty} x_i(t) = \infty$ when $x(0) \in \kappa$. Therefore $\xi = f(\kappa)$ is a degenerate vertex FP. □

6   Discussion

While Boolean networks simplify the vast complexity of cellular networks by the discretization of the phase space, switching networks admit, in addition, a finite, computable combinatorialization of the parameter space. This combinatorialization is represented as a parameter graph, where for each node in the parameter graph the dynamics in phase space are represented by a Morse graph summarizing the decomposition of the recurrent dynamics. This information is summarized as a queryable database of *Dynamic Signatures Generated by Regulatory Networks (DSGRN)* [5–8,13,17] that enables us to ask questions of biological relevance, such as the maximum number of fixed points a network can support.

The DSGRN database permits a search over dynamics (i.e equilibria, bistability, hysteresis) compatible with experimentally observed qualitative dynamics, and can assess the prevalence of a dynamical behavior in the parameter space. A natural question is how one compares the coarse combinatorial description of the dynamics in terms of Morse graphs with experimental data. Importantly, the focus shifts from matching the experimental data by trajectories generated by the model to rejecting models that are incompatible with the data. If the proposed network does not admit any parameters where the observed dynamics are present, we reject that network as a plausible model. Additionally, as we have shown recently [8] by comparing orders in sequences of maxima and minima of the experimental time series with switching systems predictions, we can reject more potential models that do not produce the observed sequences.

In this paper we have extended this combinatorial approach to a broader class of models, switching systems with regulated degradation (RD systems). This class contains, as border cases, the traditional switching systems, as well as the linearization of systems that model post-transcriptional modification of proteins (PTM systems). We show that RD systems admit state transition graphs and Morse graphs, thus allowing discretization of the phase space. We then show how to construct parameter graph for systems that mix PTM equations and switching equations. Lastly, we compare the dynamics between a switching system and a mixed PTM-switching system. We observe that the mixed systems admit only a subset of the dynamics exhibited by the corresponding switching system. Moreover, some of these dynamics are not physically meaningful, since PTM equations are not dissipative and so trajectories can tend to infinity. Therefore, some single element recurrent sets in the state transition graphs of a mixed PTM-switching system do not correspond to fixed points of the underlying differential equation.

We leave the general RD case, where individual equations contain both types of finite-valued nonlinearities for future work. The difficulty in constructing the parameter graph for the general RD system is purely technical, and not conceptual; one has to enumerate all possible relationships between values of these nonlinearities and output thresholds of the corresponding regulatory network node. This is beyond the scope of the present paper.

References

1. Albert, R., Collins, J.J., Glass, L.: Introduction to focus issue: quantitative approaches to genetic networks. Chaos **23**(2), 025001 (2013)
2. Berenguier, D., et al.: Dynamical modeling and analysis of large cellular regulatory networks. Chaos **23**(2), 025114 (2013)
3. Chaves, M., Albert, R.: Studying the effect of cell division on expression patterns of the segment polarity genes. J. R. Soc. Interface **5**(S1), S71–S84 (2008)
4. Chaves, M., Albert, R., Sontag, E.D.: Robustness and fragility of boolean models for genetic regulatory networks. J. Theor. Biol. **235**(3), 431–449 (2005)
5. Crawford-Kahrl, P., Cummins, B., Gedeon, T.: Comparison of combinatorial signatures of global network dynamics generated by two classes of ODE models. SIAM J. Appl. Dyn. Sys. **18**(1), 418–457 (2019)
6. Cummins, B., Gedeon, T., Harker, S., Mischaikow, K.: Model rejection and parameter reduction via time series. SIAM J. Appl. Dyn. Sys. **17**(2), 1589–1616 (2018)
7. Cummins, B., Gedeon, T., Harker, S., Mischaikow, K.: Database of dynamic signatures generated by regulatory networks (DSGRN). In: Feret, J., Koeppl, H. (eds.) CMSB 2017. LNCS, vol. 10545, pp. 300–308. Springer, Cham (2017). https://doi.org/10.1007/978-3-319-67471-1_19
8. Cummins, B., Gedeon, T., Harker, S., Mischaikow, K.: Model rejection and parameter reduction via time series. arXiv, 1706.04234 http://arxiv.org/abs/1706.04234 (2017)
9. de Jong, H.: Modeling and simulation of genetic regulatory systems: a literature review. J. Comput. Biol. **9**(1), 67–103 (2002)
10. Edwards, R.: Analysis of continuous-time switching networks. Physica D: Nonlinear Phenom. **146**, 165–199 (2000)

11. Edwards, R., Ironi, L.: Periodic solutions of gene networks with steep sigmoidal regulatory functions. Physica D: Nonlinear Phenom. **282**, 1–15 (2014)
12. Farcot, E.: Geometric properties of a class of piecewise affine biological network models. J. Math. Biol. **52**(3), 373–418 (2006)
13. Gedeon, T., Harker, S., Kokubu, H., Mischaikow, K., Oka, H.: Global dynamics for steep sigmoidal nonlinearities in two dimensions. Physica D **339**, 18–38 (2017)
14. Glass, L., Kauffman, S.A.: Co-operative components, spatial localization and oscillatory cellular dynamics. J. Theor. Biol. **34**(2), 219–37 (1972)
15. Glass, L., Kauffman, S.A.: The logical analysis of continuous, non-linear biochemical control networks. J. Theor. Biol. **39**(1), 103–29 (1973)
16. Gouzé, J.L., Sari, T.: A class of piecewise linear differential equations arising in biological models. Dyn. Syst. **17**(4), 299–316 (2002)
17. Huttinga, Z., Cummins, B., Gedeon, T., Mischaikow, K.: Global dynamics for switching systems and their extensions by linear differential equations. Physica D **367**, 19–37 (2018)
18. Luo, C., Wang, X.: Dynamics of random boolean networks under fully asynchronous stochastic update based on linear representation. PLoS One **8**(6), e66491 (2013)
19. Mestl, T., Plahte, E., Omholt, S.W.: A mathematical framework for describing and analysing gene regulatory networks. J. Theor. Biol. **176**, 291–300 (1995)
20. Pauleve, L., Richard, A.: Static analysis of boolean networks based on interaction graphs: a survey. Electron. Notes Theor. Comput. Sci. **284**, 93–104 (2012)
21. Thomas, R., Thieffry, D., Kaufman, M.: Dynamical behaviour of biological regulatory networks-i. biological role of feedback loops and practical use of the concept of the loop-characteristic state. Bull. Math. Biol. **57**(2), 247–76 (1995)
22. Tournier, L., Chaves, M.: Uncovering operational interactions in genetic networks using asynchronous boolean dynamics. J. Theor. Biol **260**(2), 196–209 (2009)

Reactive Models for Biological Regulatory Networks

Daniel Figueiredo[1]([✉]) and Luís Soares Barbosa[2]

[1] CIDMA, University of Aveiro, Aveiro, Portugal
daniel.figueiredo@ua.pt
[2] HASLab INESC TEC and QuantaLab, University of Minho, Braga, Portugal

Abstract. A reactive model, as studied by D. Gabbay and his collaborators, can be regarded as a graph whose set of edges may be altered whenever one of them is crossed. In this paper we show how reactive models can describe biological regulatory networks and compare them to Boolean networks and piecewise-linear models, which are some of the most common kinds of models used nowadays. In particular, we show that, with respect to the identification of steady states, reactive Boolean networks lie between piecewise linear models and the usual, plain Boolean networks. We also show this ability is preserved by a suitable notion of bisimulation, and, therefore, by network minimisation.

Keywords: Biological regulatory networks · Switch graphs · Reactivity

1 Introduction

Biochemical processes occurring within cells are abstracted into the concept of biological regulatory networks. In general, such networks capture the cell dynamics, expressed as the concentration of each component (typically proteins, and other nucleotides), which are directed by the biochemical reactions occurring between them. This process is generally regulated by the DNA through the transcription of mRNA.

Example 1. An example of a "cascade" [6] is depicted in Fig. 1. This illustrates a simple and partial biological regulatory network organised as a sequence where each node induces the production of the following one.

One of the main goals of studying biological regulatory networks is the identification of *steady states* which somehow represent the "way of working" of a cell. Different steady states can be associated, for example, to the differentiated cells of a living organism sharing the same DNA.

Models for biological regulatory networks fall in one of two major classes [7]: quantitative – where the exact concentration of each component in a cell is given; or *qualitative* – focussed on the overall dynamics of a system, and classifying the

M. Chaves and M. A. Martins (Eds.): MLCSB 2018, LNCS 11415, pp. 74–88, 2019.
https://doi.org/10.1007/978-3-030-19432-1_5

$$\boxed{\text{RIP1ub}} \longrightarrow \boxed{\text{IKK}} \longrightarrow \boxed{\text{NFkB}}$$

Fig. 1. A example of a "cascade".

concentration of a component in qualitative terms, such as "high" or "low". Quantitative models are, of course, more precise, but also harder to manipulate. A "hybrid" approach to this problem, in which a qualitative model is used to explain the "big picture" and quantitative models are applied in a second stage to study the detailed dynamics of a behavioural region [5] is a typical compromise.

This paper introduces a new qualitative model into the picture—*reactive Boolean networks* which seems particularly interesting to help in the quest for steady states, a major issue in the area. The model builds on the notion of *reactivity* [11] which proved successful in both fundamental research in modal logic [10,12,14] and software engineering applications [1,3,9].

Outline. Section 2 provides the background for this piece of research by briefly revisiting the most common models for biological regulatory networks. The notion of reactivity and the associated formal structures, namely, switch graphs and reactive frames are discussed in Sect. 3. Section 4 contains the paper main contribution, introducing reactive Boolean networks and stating some of the relevant properties. Finally, Sect. 5 concludes and identifies some directions for future work.

2   Biological Regulatory Networks

Several modelling approaches have been proposed to formally characterise biological regulatory networks [7]. This background section revisits some of them to set the scene for our own proposal discussed in Sect. 4. In general, models for biological regulatory networks consider components $i = 1, \ldots, n$, representing e.g. protein, genes, mRNA, and variables $x_1, \ldots, x_n$ corresponding to the concentration or level of expression of the respective component.

Ordinary Differential Equations and Piecewise Linear Models. The classical quantitative models resort to ordinary differential equations. The interaction between components is captured by sigmoid expressions embedded in differential equations. Either positive or negative regulations of a component i over a component j (meaning that component i respectively induces or inhibits the production/activation of component j) may be considered. This is achieved through a quantitative representation of the system. One can consider variables x_i which are related to each component i. A variable x_i represent the concentration of the component i in the system and we can use them to represent the (either positive or negative) regulations occurring between these components. This is accomplished *via* the introduction of a sigmoid function depending on each x_j in function f, in the context of the differential equation $x_i = f(x_1, \ldots, x_n)$ describing the concentration of component i.

Several classes of sigmoid functions can be chosen. A common alternative takes the form of a fraction $s^+(x; \theta, n) = \dfrac{x^n}{\theta^n + x^n}$ for positive regulations and $s^-(x; \theta, n) = 1 - s^+(x; \theta, n) = \dfrac{\theta^n}{\theta^n + x^n}$ for negative ones.

In this context, a model is obtained by considering a system with ODEs with the form $x_i = F_i(x_1, \ldots, x_n) - \gamma_i x_i$, where each $F_i(x_1, \ldots, x_n)$ is obtained as sums and products of the referred sigmoid functions and reflects how the interaction between components affect the production/expression of i. Above θ is a threshold determining the concentration of x needed to effectively regulate the target component and n determines how abrupt this regulation changes from almost inexistent to effective. Also, γ_i is a constant representing the degradation rate of the component i. All thresholds θ, n in the sigmoid functions and γ are usually estimated using suitable methods.

As one would expect, this sort of models containing non linear differential equations are hard to study analytically, but used in a number of contexts to simulate and predict the answer of a biological system. To overcome the challenge of solving a system of non linear ODEs, a more manageable alternative divides the entire state space into a finite number of domains and studies each one in its own. On a second stage the different domains are integrated and the general dynamics of the system recovered. Such models are called *piecewise linear* (PWL).

In practice, to obtain a piecewise linear model from a system of differential equations specifying a biological regulatory network, one ignores the estimated value for n and assumes that $n \to +\infty$, *i.e.*

$$\frac{x^n}{\theta^n + x^n} \xrightarrow{n \to +\infty} \begin{cases} 1, & \text{if } x > \theta \\ \frac{1}{2}, & \text{if } x = \theta \\ 0, & \text{if } x < \theta \end{cases}$$

Thus, the state space is divided into two different domains ($x < \theta$ and $x > \theta$) and a boundary $x = \theta$. Given a specific model, this technique is applied to each one of these sigmoid functions in order to split the state space into several regions with a linear differential equation expressing the state trajectory within each one.

Example 2. Consider the following system of differential equations:

$$\begin{cases} x' = 5\dfrac{x^2}{x^2 + 2^2} \cdot \dfrac{2^2}{y^2 + 2^2} - x \\ y' = 3\dfrac{x^2}{x^2 + 4^2} - y \end{cases}$$

Making $n \to +\infty$ leads to

$\begin{cases} x' = -x \\ y' = -y \\ \quad x < 2 \\ \quad 2 < y \end{cases}$	$\begin{cases} x' = -x \\ y' = -y \\ 2 < x < 4 \\ \quad 2 < y \end{cases}$	$\begin{cases} x' = -x \\ y' = 3 - y \\ \quad 4 < x \\ \quad 2 < y \end{cases}$
$\begin{cases} x' = -x \\ y' = -y \\ \quad x < 2 \\ \quad y < 2 \end{cases}$	$\begin{cases} x' = 5 - x \\ y' = -y \\ 2 < x < 4 \\ \quad y < 2 \end{cases}$	$\begin{cases} x' = 5 - x \\ y' = 3 - y \\ \quad 4 < x \\ \quad y < 2 \end{cases}$

Analytically, a steady state can be identified at point $(0,0)$, and an orbit which asymptotically converges to $(4,2)$.

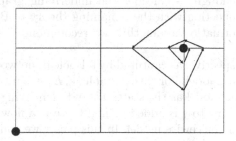

Fig. 2. Steady states identified in Example 2.

Formally, it is important to mention that, in general, a more careful study at the domains should be performed. However, we do not focus much on this issue for now because we will only consider PWL models where no such care is required.

In the context of a differential equation, we say that a *flow* is a trajectory obtained from a initial state which is ruled by the differential equations. In the context of a PWL model, we give a similar meaning to this term but we admit a flow to be the concatenation of several usual flows, obtained within different domains, regarding that this concatenated flow still is continuous. For instance, in Fig. 2, a flow in a PWL model is illustrated by a solid line. The orientation represented by the arrows describes the evolution of the flow along time.

Boolean Networks. A Boolean network (BN) is another kind of model to describe the dynamics of biological regulatory networks. This kind of model considers the concentrations x_i as Boolean variables assuming, in practice, $x_i = 0$ if the concentration of i is "low" (bellow some threshold) and $x_i = 1$ if the concentration of i is "high" (above some threshold).

In a BN model, each variable x_i is regulated by a Boolean function $f_i(x_1, \ldots, x_n)$ built from Boolean operators, combined in a general function

$f = (f_1, \ldots, f_n)$. From this function, a graph (Boolean network) (V, E) can be obtained. At this point, two paradigms can be considered: synchronous and asynchronous. For the synchronous approach, the BN is obtained as follows: given the set of components A with $|A| = n$, $V = \{0, 1\}^n$ and a directed edge $(a, b) \in E$ if $b = f(a)$. In this way, each vertex in V admits exactly one output. For the asynchronous approach V is obtained as before but a directed edge $(a, b) \in E$ if there is some index i such that $b_i = f(a)_i \neq a_i$ and $b_j = a_j$ for every other indexes $j \neq i$. In this way, only the value of a single variable is updated at each step. Thus, we admit networks where several edges can have the same vertex as tail and where vertices with no outgoing edges are allowed. The asynchronous approach was proposed and is often used since it reflects the dynamics of the system in a more realistic way.

For our approach we do not need to consider these theoretical concepts regarding BNs in detail, since our network will be built using a different procedure. Thus, it is enough to consider the underlying graphs as BN models, although a wider connection with the remaining theory of BNs could be established. For more information about this, we recommend the reader to consult [13] for more details.

A qualitative perspective, as captured by a Boolean network, can be obtained from a piecewise linear model. For this, a graph (V, E) is built taking the domains of the piecewise linear model as the states V, and identifying the edges with the flows, *i.e.* an edge from i to j is added to E if there is a flow from the domain i to the domain j in the piecewise model. In this paper we consider asynchronous dynamics, meaning, in practice, that we only admit "adjacent" domains to be connected.

As referred, in BNs, variables can take values 1 or 0 depending on the concentration/level of expression of the corresponding protein/gene being, respectively, high or low (above or bellow a threshold θ). Since in PWL models we consider thresholds to split the state space, these Boolean variables are then used to identify each domain. However, in order to be able to fully accomplish this, sometimes more than two values must be considered. For instance, a third value may be added to cater for intermediate states. This will be the case in the next example which illustrate how we need to consider three possible values for x—0, 1 and 2—for low, medium and high concentration, respectively.

Example 3. Figure 3 shows a BN corresponding to the biological regulatory network described in Example 2. We can think of it as a graph where the vertices have the form ab where a and b represent the level of the corresponding component. We note that we must consider three values for the variable related to a since in the PWL model two thresholds are considered for the variable x. For instance state 00 represents the state where both components have low concentration and 10 is the one whose first component has a medium concentration and the second component has a low concentration. Recalling its piecewise linear model, we note that each vertex corresponds to a domain and the edges are introduced according to the flows within each domain. For instance, there is an edge from 11 to 01 because, in the corresponding PWL model it is possible to attain the boundary between the domain represented by 01 and 11 with a flow whose initial state is $(x, y) = (\frac{5}{2}, \frac{7}{2})$, which is within the domain denoted by 11.

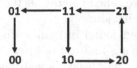

Fig. 3. Example of a Boolean network.

Steady states in a BN are identified by strongly connected components with no outgoing edges [7], which is called an *attractor* in the literature. Given a graph (V, E), a strongly connected component is a set of vertices X such that, for every $x, y \in X$, there exists a directed path between x and y which only contains vertices in X.

The BN depicted in Example 3 admits two strongly connected components – $\{00\}$ and $\{11, 21, 20, 10\}$ – but only one of them is an attractor and, thus, represents a steady state, since $\{11, 21, 20, 10\}$ admits an outgoing edge from 11 to 01. Actually, it is usual to lose information when going from a piecewise linear model to a Boolean network. In general, it is well known that, for a system S,

$$SteadySt([\![S]\!]_{BN}) \subseteq SteadySt([\![S]\!]_{PWL}) \subseteq SteadySt([\![S]\!]_{ODE})$$

where $[\![S]\!]_M$ refers to the representation of system S in model M and $SteadySt$ retrieves the set of steady states in a model.

As expected, simpler models tend to lose information about the steady states of the system. Our proposal, discussed in the sequel tries to partially correct this problem.

3   Switch Graphs and Reactive Frames

As mentioned in Sect. 1, the model proposed in this paper borrows from modal logic research the notion of *reactive frame*. Modal logics [4] are logics in which formulæ are interpreted over a graph of semantic universes interconnected through an accessibility relation. The former may represent *e.g.* temporal instants, deontic contexts or epistemic states. In all cases truth is relative rather than universal as in classical logics which assume just one universe of interpretation. Modal operators—□ and ◊—provide a universal, respectively, existential, quantification over the accessible universes from the current point of evaluation.

Typically this accessibility relation (*i.e.* the underlying graph) is fixed. A *reactive frame*, however, is a graph whose structure can vary over time. Such graphs have been used as semantic models for some classes of modal logics in which the way semantic universes are interconnected can be modified on the fly.

The idea can be traced back to Johan van Benthem seminal paper on the so-called sabotage logic [3] in which an edge is deleted after being taken (therefore preventing its subsequent use). In another variant described by Areces *et al.* [1], edges are not deleted but their direction swapped. This sort of logics [2] and the underlying dynamic graphs are called *reactive* as they capture structural changes under reaction to previous behaviour.

Reactive behaviour is often found in games. For instance, castling, a chess special move involving the King and a Rook, can only be performed if none of the pieces which take part has been moved before. Therefore, in identical configurations of the table, different moves can be possible. A similar situation involves another chess move called *en passant*, which allows a player to capture a pawn from the opponent with one of his pawns whenever the opponent moves his pawn two squares in front and it becomes laterally adjacent to an enemy pawn. However, if the opponent only moves his pawn a square in front, *en passant* becomes illegal.

3.1 Switch Graphs

Syntactically, such moving structures are represented by *switch graphs* [10,12], which add higher-level edges to the usual graph structure. These higher-level edges connect basic edges, also called "0-level" edges, which are the ones which are eventually crossed, to other higher-level edges according to the following definition.

Definition 1. *Given a set of nodes* W*, a switch graph is a pair* (W, S) *where* $S = \bigcup_{n \geq 0} S_n$ *such that:*

- $S_0 \subseteq W \times W$, i.e. *the usual relation between nodes,*
- *and, for* $n \geq 1$, $S_n \subseteq S_0 \times S_{n-1} \times \{\circ, \bullet\}$.

A higher-level edge $(d, e, *)$ will either inhibit or activate its target edge e whenever the source edge d is crossed, depending on the value of annotation $*$. Target edge e will be inhibited if $* = \circ$ or activated if $* = \bullet$. In the graphical representation of a switch graph, as shown in Fig. 4, inhibitor edges are depicted as white headed arrows, while black headed arrows represent activator edges.

A switch graph is configured through an *instantiation* function $I : S \to \{0, 1\}$ which marks each edge as inhibited or active depending on $I(s) = 0$ or $I(s) = 1$, respectively. The former (respectively, latter) edges are depicted as dashed (respectively, full) arrows. Note that inhibited edges cannot be crossed, and they can neither activate nor inhibit other edges. Moreover, only 0-level edges can be crossed: if one such edge x is crossed, all active higher-level edges with source in x, i.e. $(x, e, *)$ will fire and activate/inhibit the respective target edge e.

Example 4. Figure 4 depicts a switch graph (W, S) with $W = \{w\}$ and

$$S = \{(w, w), ((w, w), (w, w), \circ), ((w, w), ((w, w), (w, w), \circ)), \bullet)\}$$

For simplicity, we define $e_1 = ((w,w),((w,w),\circ)),\bullet)$ and $e_2 = ((w,w),(w,w),\circ)$. The initial instantiation I_0 is such that $I_0(w,w) = 1$ (the edge (w,w) can be crossed), $I_0(e_2) = 0$ (meaning that it is inhibited) and $I_0(e_1) = 1$ (therefore, activated and ready to activate e_2, the pointed edge whenever (w,w) is crossed).

Therefore, starting from w, the edge (w,w) can be crossed (since it is active) and this causes the higher-level edge e_1 to fire and activate e_2. e_2 has no effect since it was initially inhibited when (w,w) was crossed. One can then cross (w,w) again. Now, e_1 acts but has no effect, since e_2 is already active, while e_2 acts and inhibits (w,w). Hence, (w,w) can no longer be crossed. This switch graph shows an example of counter which can "count" only twice.

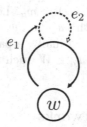

Fig. 4. Example of a switch graph.

3.2   From Switch Graphs to Reactive Frames

A switch graph can be encoded by the set of paths it generates, as shown in the following example.

Example 5. The set of paths Δ corresponding to the switch graph shown in Fig. 5 is generated as follows. Starting at w_1 we can move to w_3 but not from there since the edge (w_3, w_4) was deleted on crossing (w_1, w_3). Thus, $(w_1), (w_1, w_3) \in \Delta$. Starting at w_2 we can move to w_3 and afterwards to w_4 (since the edge (w_3, w_4) was preserved). Thus, $(w_2), (w_2, w_3), (w_2, w_3, w_4) \in \Delta$. Starting at w_3 there is a move to w_4, from where no other move is possible. Therefore, $\Delta = \{(w_1), (w_1, w_3), (w_2), (w_2, w_3), (w_2, w_3, w_4), (w_3), (w_3, w_4), (w_4)\}$.

Paths are used to define *reactive frames*—a semantic model for switch graphs upon which suitable (reactive) modal logics are defined. We will not develop such logics in this paper, the interested reader is referred to [10,12] for an extensive account.

Definition 2. *Consider $W \neq \emptyset$ and let $\Delta \subseteq W^*$ be a nonempty set of finite paths. (W, Δ) is a reactive frame if:*

- *$(w) \in \Delta$ for any $w \in W$*
- *$\forall n \geq 1, (w_1, \ldots, w_n, w_{n+1}) \in \Delta$ implies $(w_1, \ldots, w_n) \in \Delta$.*

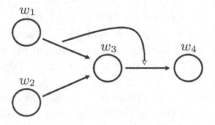

Fig. 5. Another switch graph.

Let us end this section, fixing some notation. As usual, we denote by W^* the set of all non-empty finite sequences (*paths*) over set W, and define function $t : W^* \to W$ by $t(\lambda) = t(w_1, \ldots, w_n) = w_n$. Let $\lambda = (w_1, \ldots, w_n)$ be a path. Notation λw abbreviates the path $(w_1, \ldots, w_n, w)$. Similarly, if no ambiguity arises, w_1 stands for the path (w_1). Finally, a path γ *extends* or *is an extension* of a path λ is there exist $w_0, \ldots, w_n \in W$ such that $\gamma = \lambda w_0 \ldots w_n$. Of course, every path is an extension of itself.

4   Reactive Boolean Networks

Switch graphs provide an interesting alternative to represent biological regulatory networks, building on the corresponding Boolean networks. As discussed below, this new model which we propose to call *reactive Boolean networks* (RBN) has a number of interesting properties namely in what concerns the identification of steady states and their preservation under model minimisation.

In the next definition, we consider a PWL model and the corresponding Boolean network obtained from it. There, when we mention a vertex x of the Boolean network we also mention the corresponding domain of the PWL model and *vice-versa*.

Definition 3. *Given a PWL model M whose corresponding Boolean network is N, a reactive boolean network is a two-level switch graph (W, S) where $(W, S_0) = N$ and S_1 is obtained according to the following rules:*

1. *For any domain k of M such that $u = (j, k) \in S_0$, we have $(v, u, \circ) \in S_1$ with $v = (i, j)$ if a flow which enters in region j via the boundary between regions i and j never leaves it via the boundary between regions j and k.*
2. *For each $(v, u, \circ) \in S_1$ with $u = (j, k)$ and $v = (i, j)$, then $(w, u, \bullet) \in S_1$ if there exists $w = (l, j) \in S_0$ for some region l of M such that there is a flow entering in region j via the boundary between regions l and j and leaving it via the boundary between j and k.*

Moreover, since (W, S) is a two-level switch graph, then $S_n = \emptyset$, for $n > 1$.

We only define Reactive Boolean networks for two-level switch graphs. In fact, this definitions could be generalized to embed higher-level edges but we believe that the benefits would not be worth the additional computational cost. Then, we leave the this generalization for future work.

In practice, this new kind of models can temporarily deleted from the state transition graph edges that would represent non-realistic behaviours. In practice, and since we can compute the flow given by a linear differential equation and an initial state, the inclusion of an edge of the type $((i,j),(j,k),\circ) \in S_1$ means that is not possible to obtain, in the PWL model, a flow with initial state in the region i that enters in the region j and leads us to region k.

4.1   Recovering Attractors

As mentioned before, it is well known that, in general:

$$SteadySt(\llbracket S \rrbracket_{BN}) \subseteq SteadySt(\llbracket S \rrbracket_{PWL})$$

with equality failing for multiple examples. Reactive Boolean networks, on the other hand, in general can increase the number of steady states that can be identified when comparing to Boolean networks, introducing a further level in this inequality:

$$SteadySt(\llbracket S \rrbracket_{BN}) \subseteq SteadySt(\llbracket S \rrbracket_{RBN}) \subseteq SteadySt(\llbracket S \rrbracket_{PWL})$$

In the context of reactive Boolean networks, steady states are also identified by *atractors*, whose definition is revised as follows.

Definition 4. *Given a reactive Boolean network (W, S) whose set of paths is Δ, a set $V \subseteq W$ forms a* strongly connected component relatively to a path $\lambda \in \Delta$ *(SCC_λ) if for any $v \in V$ and any path $\rho \in \Delta$ which extends λ, there exists $\gamma \in \Delta$ such that $t(\gamma) = v$ and γ extends ρ.*

Proposition 1. *If V is a SCC_λ, it is always possible to find a path between two states $u, v \in V$ after a reconfiguration on the edges, induced by the path λ.*

Proof. From the definition, for all extensions ρ of λ, one can find λ an extension of ρ such that $t(\gamma_u) = u$. Again, by definition, and since γ_u is itself extends λ, it is possible to find γ_v which extends γ_u and such that $t(\gamma_v) = v$.   □

Definition 5. *Given a reactive Boolean network (W, S) whose set of paths is Δ, a set $V \subseteq W$ is an* attractor *if it is a SCC_λ, for some path λ, and every path γ extending λ verifies $t(\gamma) \in V$.*

This definition extends the notions of SCC and attractor, which are defined for regular graphs (and BNs) to switch graphs (and RBNs). Informally, this means that a set V is an attractor if there is a path γ such that, after walking along it, we can always find a path between any two states of V and there is not any path guiding us to a final state outside the set V (*i.e.*, the usual definition for attractor in a usual graph).

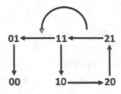

Fig. 6. A reactive Boolean network.

Example 6. Recall Examples 2 and 3, and take the Boolean network introduced then as the first stage (W, S_0) of a reactive Boolean network. The piecewise linear model in Example 2 generates S_1 which turns out to be a singleton relation $\{((21, 11), (11, 01), \circ)\}$. The whole net is depicted in Fig. 6.

For this reactive network, given 21 as the initial state, we obtain the following set of paths: $\{(21), (21, 11), (21, 11, 10), (21, 11, 10, 20), (21, 11, 10, 20, 21), \ldots\}$. Therefore, according to Definition 5, $\{11, 10, 20, 21\}$ is an attractor for this reactive Boolean network. Similarly, taking 00 as the initial state, $\{00\}$ emerges as an attractor as well.

Definition 6. *Consider a BN or RBN model (V, E) and a RBN model (V', E'). Given an attractor $A \subseteq V$ of the model (V, E), we say that it is signaled by the model (V', E') if there is an algorithm that allows us to obtain a set $B \subseteq V'$ from A in an unambiguous way and such that B is an attractor of (V', E').*

Using other words, if a model "A" signals the attractors of another model "B", we can recover all attractors of the model "B" from the attractors of the model "A".

Proposition 2. *Given a piecewise linear model, the corresponding reactive Boolean network identifies, in general, a larger set of attractors than the simpler Boolean network. Moreover, all attractors of a BN are signalled in the corresponding RBN.*

Proof. The fact that, in general, a RBN a larger set of attractors from than a BN was already shown in Example 6. Note that we were able to recover the attractor 00 as well as the converging cyclic behaviour of the piecewise linear model that the BN was not able to signal.

Now, consider a BN with an attractor V. If $|V| = 1$, then V is also attractor of the corresponding *RBN*. Otherwise, let $\lambda = (v)$ be a path with $v \in V$. Since V is an attractor in a BN, all extensions of λ terminate at elements of V. Consider the following algorithmic procedure: Choose $v \in V$ for which it is possible to consider an extension γ of λ such that it is no more possible to extend it to path ρ where $t(\rho) = v$. If such a path exists, update λ to γ and V to $V \setminus \{v\}$. Repeat this process while it is possible to choose such a v. Note that, since V and S are finite (*i.e.* there is a finite number of configurations for the S_0 edges), this algorithm terminates. Note that, after this process, for any $v \in V$ such that there is an extension γ_v of λ such that $t(\gamma_v) = v$, and for every $w \in V$, there is an extension ρ_w of γ_v such that $t(\rho_w) = w$. This proves that V is a SCC_λ and, since each extension γ of λ is such that $t(\gamma) \in V$, the RBN signals an attractor.   □

Proposition 3. *All attractors in a RBN are steady states of the corresponding piecewise linear model.*

Proof. A steady state of a PWL can be found as either an invariant region or cyclic behaviour which assymptoticaly converges to a point or orbit. Let V be an attractor in a RBN, and consider a region T resulting from the union of every domain represented by $i \in V$. Since V is a SCC_λ for some path λ, this means that there is a flow in the piecewise linear model which makes impossible to leave region T. This means that there exists an invariant subregion T' of T and, therefore, it contains a steady state.   □

4.2   Bisimulation and Minimisation

In a previous publication [9], the authors defined a notion of bisimulation for reactive frames and proved a Hennessy-Milner like theorem stating the equivalence between bisimilarity (*i.e.* the existence of a bisimulation relating two nodes in a frame), and logical validity (*i.e.* the fact that both nodes satisfy exactly the same set of formulæ expressed in a suitable modal logic). From the modelling point of view adopted in this paper, bisimulation is a crucial tool to reduce the size of a reactive frame while keeping the behaviour it may induce, thus increasing the performance of any automatic analysis tool operating over reactive frames. Some examples of application of bisimulation to non reactive models can be found in [8,15]. Bisimulations can be also used for other purposes in a biological context: for example, in [8], attractors of a Boolean models are highlighted using bisimulations.

Definition 7. *Let (W, Δ) and (W', Δ') be two reactive frames. A relation $S \subseteq \Delta \times \Delta'$ is a bisimulation if and only if $\forall \lambda \in \Delta, \forall \lambda' \in \Delta'$ such that $(\lambda, \lambda') \in S$:*

R-zig: $\forall w \in W(\lambda w \in \Delta \Rightarrow \exists w' \in W', \lambda' w' \in \Delta'$ *such that* $(\lambda w, \lambda' w') \in S)$
R-zag: $\forall w' \in W'(\lambda' w' \in \Delta' \Rightarrow \exists w \in W, \lambda w \in \Delta$ *such that* $(\lambda w, \lambda' w') \in S)$
P-zig: $\forall \gamma \in \Delta(t(\lambda) = t(\gamma) \Rightarrow \exists \gamma' \in \Delta'(t(\lambda') = t(\gamma')$ *and* $(\gamma, \gamma') \in S))$
P-zag: $\forall \gamma' \in \Delta'(t(\lambda') = t(\gamma') \Rightarrow \exists \gamma \in \Delta(t(\lambda) = t(\gamma)$ *and* $(\gamma, \gamma') \in S))$

Example 7. Figure 7 depicts two switch graphs which induce bisimilar reactive models. In fact, we can verify that the following relation is a bisimulation:

$$\{((w_1), (v_1)), ((w_1, w_2), (v_1, v_2)), ((w_2), (v_2)), ((w_2, w_3), (v_2, v_2)), ((w_4), (v_1)),$$
$$((w_4, w_3), (v_1, v_2)), ((w_3), (v_2)), ((w_3, w_2), (v_2, v_2))\}$$

Bisimulation can also easily be formulated for switch graphs, and therefore for reactive Boolean networks, as follows.

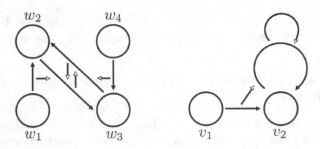

Fig. 7. Two switch graphs whose corresponding reactive frames are bisimilar.

Definition 8. *Given two switch graphs* (W, S), (W', S') *whose reactive frames are* (W, Δ) *and* (W', Δ'), *and an equivalence relation* $\mathcal{R} \subseteq W \times W'$, *we say that a relation* $\mathcal{B} \subseteq \Delta \times \Delta'$ *is induced by* R *when* $\mathcal{B}$ *is such that:*

- $(w, w') \in \mathcal{R} \Leftrightarrow ((w), (w')) \in \mathcal{B}$ *for every* $w \in W$, $w' \in W'$.
- *Let* $\lambda \in \Delta$ *and* $\lambda' \in \Delta'$ *be such that* $(\lambda, \lambda') \in \mathcal{B}$. *For every* $w \in W$ *and* $w' \in W'$ *such that* $\lambda w \in \Delta$ *and* $\lambda' w' \in \Delta'$, *we have* $\lambda w \mathcal{B} \lambda' w'$ *iff* $(w, w') \in \mathcal{R}$.

Moreover, we say that $\mathcal{R}$ *is a bisimulation iff the induced relation* $\mathcal{B}$ *is a bisimulation for reactive models.*

From a model (W, S), we can soundly obtain a reduced model (W', S') if there exists a bisimulation R verifying $\forall w \in W \exists w', wRw'$. In this case, we say R is *total*. Our final result states that bisimulation preserve attractors.

Lemma 1. *Let* (W, S) *and* (W', S') *be two switch graphs,* (W, Δ) *and* (W', Δ') *be corresponding reactive frames. Let* $\mathcal{R}$ *be a total bisimulation between* (W, S) *and* (W', S') *and* $\mathcal{B}$ *be the induced relation from* $\mathcal{R}$. *Then* $\mathcal{B}$ *is total whenever* $\mathcal{R}$ *is total.*

Proof. We prove this lemma by induction over paths.

Let $\lambda \in \Delta$ be a path. If $\lambda = (w)$, for some $w \in W$, then, since R is total, there is $w' \in W'$ such that $(w, w') \in \mathcal{R}$ and, therefore $(w)\mathcal{B}(w')$.

Let us now consider a path $\gamma = \lambda w$ for some $w \in W$ and $\lambda \in \Delta$. Then, by induction hypothesis, there are $\lambda' \in \Delta'$ such that $\lambda \mathcal{B} \lambda'$. Then, since $\mathcal{B}$ is a bisimulation and by definition, $\exists w' \in W'$ such that $\lambda w \mathcal{B} \lambda' w'$. □

Proposition 4. *Let* (W, S) *and* (W', S') *be two bisimilar switch graphs. Each attractor of* (W, S) *is signaled by some attractor of* (W', S').

Proof. Let (W, S) and (W', S') be two switch graphs whose corresponding reactive frames are (W, Δ) and (W', Δ'), respectively. Let also $R \subseteq W \times W'$ be a total bisimulation and T the corresponding bisimulation for reactive frames.

Consider A, an attractor of (W, S). Thus, there is some path $\lambda \in \Delta$ such that A is a SCC_λ. According to the previous lemma and since $\mathcal{R}$ is total, then $\mathcal{B}$ is also total. Then, by definition, there is $\lambda' \in \Delta$ such that $(\lambda, \lambda') \in T$.

Let $\bar{B} = \{t(\gamma') : \gamma' \in \Delta'$ be an extension of $\lambda'\}$. By the definition of $\bar{B}$, and using a process analogous to the one presented in the proof of Proposition 2, we obtain an attractor $B \subseteq \bar{B}$. We will show that the states of B are related with the states of A.

If $b \in B$, then it means that there is an extension γ' of λ' such that $(\gamma') = b$, i.e. $\exists w'_0, \ldots, w'_n$ such that $\gamma' = \lambda'w_0 \ldots w_n$. Since R is a bisimulation, we know that $t(\lambda)Rt(\lambda')$, $t(\lambda w_0)Rt(\lambda'w'_0)$, $\ldots$, $t(\lambda w_0 \ldots w_n)Rt(\lambda'w_0 \ldots w'_n)$, where $w_0, \ldots, w_n \in W$ are such that $w_0Rw'_0$, $\ldots$, $w_nRw'_n$. Since A is an attractor w_0, $\ldots$, $w_n \in A$. $\qquad\square$

We end this paper with some considerations about the expressibility of switch graphs when compared with regular graphs. In fact, they present similar expressibility as when can think, in some sense, in a translation of switch graphs to usual graphs: a switch graph can be seen as a regular graph where each state is a pair (x, I) where x is a state of the switch graph and I is an admissible instantiation. The accessibility relation is defined such that there is an edge between two states (x, I) and (y, J) if $I(x, y)$ is defined and it is equal to I, and I is updated to J when the edge (x, y) is crossed. Although this "translated" model is not so intuitive as the switch graph itself, it allows us to obtain a finite and usual graph from a switch graph (W, S) whenever S is finite. Therefore, it allows us to apply the already existing tools in Computer Science to study switch graphs.

5   Conclusions and Future Work

This paper proposed a new model for biological regulatory networks based on the notion of *reactivity* as introduced by Gabbay and his collaborators in the context of transition systems and their modal logics. The proposed model—*reactive Boolean networks*—is discrete and finite (therefore amenable to transformation to a plain graph representation). Similarly to the usual Boolean networks, the reactive ones provide a straightforward way to simplify piecewise linear models. However, as shown here, its ability to identify steady states overcomes usual Boolean networks. We also show that such an ability is preserved under a suitable notion of bisimulation and, therefore, under network minimisation.

There are several avenues for future work we are currently exploring. The first one consists of introducing weighted edges in the reactive network to capture either some form of uncertainty in the cell evolution or describe the consumption of contextual resources. We also intend to resort to the reactive modal logic [10] interpreted over reactive frames to formulate and verify properties of biological regulatory networks.

Acknowledgments. This work was supported by ERDF - The European Regional Development Fund through the Operational Programme for Competitiveness and Internationalisation - COMPETE 2020 Programme and by National Funds through the Portuguese funding agency, FCT - Fundação para a Ciência e a Tecnologia, within project POCI-01-0145-FEDER-030947. and project with reference UID/MAT/04106/2019 at

CIDMA. D. Figueiredo also acknowledges the support given by FCT via the PhD scholarship PD/BD/114186/2016.

The authors are also grateful to the reviewers for their useful comments and corrections.

References

1. Areces, C., Fervari, R., Hoffmann, G.: Swap logic. Log. J. IGPL **22**(2), 309–332 (2013). p. jzt030
2. Areces, C., Fervari, R., Hoffmann, G.: Relation-changing modal operators. Log. J. IGPL **23**(4), 601–627 (2015). p. jzv020
3. van Benthem, J.: An essay on sabotage and obstruction. In: Hutter, D., Stephan, W. (eds.) Mechanizing Mathematical Reasoning. LNCS (LNAI), vol. 2605, pp. 268–276. Springer, Heidelberg (2005). https://doi.org/10.1007/978-3-540-32254-2_16
4. Blackburn, P., De Rijke, M., Venema, Y.: Modal Logic: Graph. Darst, vol. 53. Cambridge University Press, Cambridge (2002)
5. Chaves, M.: Predictive analysis of dynamical systems: combining discrete and continuous formalisms. Ph.D. thesis, Gipsa-lab (2013)
6. Chaves, M., Tournier, L.: Predicting the asymptotic dynamics of large biological networks by interconnections of boolean modules. In: 2011 50th IEEE Conference on Decision and Control and European Control Conference (CDC-ECC), pp. 3026–3031. IEEE (2011)
7. De Jong, H.: Modeling and simulation of genetic regulatory systems: a literature review. J. Comput. Biol. **9**(1), 67–103 (2002)
8. Figueiredo, D.: Relating bisimulations with attractors in boolean network models. In: Botón-Fernández, M., Martín-Vide, C., Santander-Jiménez, S., Vega-Rodríguez, M. (eds.) AlCoB 2016. LNCS, vol. 9702, pp. 17–25. Springer, Cham (2016). https://doi.org/10.1007/978-3-319-38827-4_2
9. Figueiredo, D., Martins, M.A., Barbosa, L.S.: A note on reactive transitions and reo connectors. In: de Boer, F., Bonsangue, M., Rutten, J. (eds.) It's All About Coordination. LNCS, vol. 10865, pp. 57–67. Springer, Cham (2018). https://doi.org/10.1007/978-3-319-90089-6_4
10. Gabbay, D., Marcelino, S.: Global view on reactivity: switch graphs and their logics. Ann. Math. Artif. Intell. **66**(1–4), 1–32 (2012)
11. Gabbay, D.M.: Reactive Kripke Semantics. Cognitive Technologies. Springer, Heidelberg (2013). https://doi.org/10.1007/978-3-642-41389-6
12. Gabbay, D.M., Marcelino, S.: Modal logics of reactive frames. Stud. Logica **93**(2), 405–446 (2009)
13. Glass, L., Kauffman, S.A.: The logical analysis of continuous, non-linear biochemical control networks. J. Theor. Biol. **39**(1), 103–129 (1973)
14. Marcelino, S.R.T.: Modal logic for changing systems. University of London (2011)
15. Pola, G., Di Benedetto, M.D., De Santis, E.: Arenas of finite state machines. arXiv preprint arXiv:1106.0342 (2011)

Temporal Logic Based Synthesis of Experimentally Constrained Interaction Networks

Judah Goldfeder[1] and Hillel Kugler[2(✉)]

[1] Yeshiva University, New York, USA
ygoldfed@gmail.com
[2] Bar-Ilan University, Ramat-Gan, Israel
hillelk@biu.ac.il

Abstract. Synthesis methods based on formal reasoning are a powerful way to automate the process of constructing computational models of gene regulatory networks (GRNs) and increase predictive power by considering a set of consistent models that are guaranteed to satisfy known experimental data. Previously, a formal reasoning based approach enabling the synthesis and analysis of biological networks formalized using Abstract Boolean Networks (ABNs) was developed, where the precise interactions and update rules are only partially known. System dynamics can be constrained with specifications of some required behaviors, thereby providing a characterization of the set of all networks capable of reproducing given experimental observations. The synthesis method is supported by a tool, the Reasoning Engine for Interaction Networks (RE:IN). Starting with the synthesis framework supported by RE:IN, we provide translations of experimental observations to temporal logic and semantics of Abstract Boolean Networks, enabling us to use off-the-shelf model checking tools and algorithms. An initial prototype implementation we have developed demonstrates this is a gainful approach, providing speed-up gains for some benchmarks, while also opening the way to study extensions of the experimental observations specification language currently supported in RE:IN by using the rich expressive power of temporal logic.

Keywords: Gene regulatory networks · Formal verification ·
Boolean networks · Synthesis · Temporal logic

1 Introduction

Building computational models of gene regulatory networks can be an effective way to improve our understanding of a biological system, identify gaps in our knowledge, and make new predictions that can be tested experimentally. A major challenge while constructing such computational models is ensuring that a model indeed explains the known experimental data, and avoiding the implicit bias

© Springer Nature Switzerland AG 2019
M. Chaves and M. A. Martins (Eds.): MLCSB 2018, LNCS 11415, pp. 89–104, 2019.
https://doi.org/10.1007/978-3-030-19432-1_6

that may be introduced by the modeler while choosing between many consistent possible models.

Synthesis methods based on formal reasoning can help in tackling these challenges by automating the process of constructing consistent models, and by enabling to make predictions using the set of all consistent models of a specified form instead of one model that may contain some implicit assumptions. The Reasoning Engine for Interaction Networks (RE:IN) [12,35] supports such a synthesis approach by allowing for the specification of an Abstract Boolean Network (ABN), which leaves open the option to include or to omit some of the network interactions. This also gives freedom in selecting the regulation conditions, which determine the combined dynamical effect of a component's activators and repressors.

RE:IN also allows to constrain the ABN by specifying experimental constraints on system dynamics based on known experimental measurements of the biological system. The RE:IN synthesis procedure can then combine the ABN with the experimental constraints and synthesize consistent models, or prove that no consistent model exist, which is a key ability of formal verification methods, in contrast to most machine learning based inference methods (e.g. [16,26,27]) that do not prove the absence of solutions but can be very effective in identifying approximate solutions. If RE:IN proves that no consistent model exist, the user should double check the encoding of the model and also evaluate the assumptions made, refining the ABN and experimental constraints as necessary. RE:IN has been used successfully to study the pluripotency program of mouse embryonic stem cells [12] and investigate stem cell reprogramming [11]. The RE:IN tool can support a range of synthesis and analysis queries and provides a user-friendly interface for visualizing the networks, experiments and analysis results. Additional information on the tool and methods appear in [35], including application to the cell cycle in budding yeast, myeloid progenitor differentiation, and the murine cardiac gene regulatory network controlling First Heart Field and Second Heart Field differentiation. An extension to networks that can dynamically reconfigure their interactions, termed switching networks is presented in [28,32].

To synthesize consistent models, RE:IN encodes the synthesis problem using the Z3 Satisfiability Modulo Theories (SMT) solver [8] and utilizes a bounded model checking strategy to search for a consistent model or prove that no such model exists. The efficiency of SMT solvers and experience gained in studying alternative strategies for problem encoding has enabled RE:IN to tackle realistic problems of biological interest [12,35,36]. However for some of the models performance is becoming a challenge, especially as the number of experimental observations grows [11]. In this work we investigate ways to improve the running time of the synthesis methods and to eventually enable tackling larger networks with more experimental constraints than currently feasible. Towards achieving this goal we show how experimental observations in RE:IN can be encoded using temporal logic, and how ABNs can be encoded as a transition system in standard model checking tools, focusing on the NuSMV model checker [6]. This encoding allows us to use the model checking algorithms of NuSMV to tackle the synthesis problem. An additional benefit of this approach is that the expressive power of temporal logic can be utilized to extend the specification language currently supported by RE:IN, allowing to encode other interesting biological properties that can be expressed in temporal logic [14].

The remainder of the paper is structured as follows: We define how the RE:IN synthesis problem can be encoded in NuSMV and solved by standard model checking procedures. The problem consists of two parts: specification of the Abstract Boolean Network (Sect. 2), and specification of experimental constraints on this network in the form of observations from experiments (Sect. 3).

In Sect. 2 we encode the ABN using NuSMV modules to specify a transition system, with each module corresponding to a node, and the main module corresponding to the entire network. Each module specifies the connecting nodes as inputs into the module. For optional connections, the state of connectivity is modeled as state variables. Possible activation functions (patterns that define the overall effect of activators and repressors) are represented as possible transitions to the next state.

In Sect. 3 we show how the experimental constraints on the network dynamics are encoded into temporal logic. Two variants are discussed, with the trade-offs between the two explored. The first method uses Linear Temporal Logic (LTL) to constrain the network to synthesize results consistent with the observed experiments, while the second method uses Computational Tree Logic (CTL) to do the same. Once the problem is properly encoded, we run NuSMV model checking. If there are any concrete networks consistent with the experimental observations, NuSMV will provide a counter-example telling us what one such concrete network is. If no such network exists, NuSMV will prove that and report that no solution for the specification exists. Finally, related work is presented in Sect. 4.

2   Abstract Boolean Network Semantics

We now describe how Abstract Boolean Networks are encoded as a transition system, using the SMV language supported by NuSMV [6]. In RE:IN an ABN is specified in a separate model file (with a .net extension) in the form of two lists. First is a node list, containing the name of every node in the network (nodes typically correspond to genes, transcription factors or signals), every possible activation function associated with each node, and whether each node is able to be over-expressed (FE) or knocked-out (KO). An activation function (termed regulation condition in RE:IN) for a particular node is the function that determines, given the current state of activation of all nodes that interact with it, whether this node will be activated or not in the next time-step. Second is a connection list, including all connections between nodes as well as whether each connection is optional or mandatory (definite in RE:IN terminology), activating or repressing.

In our proposed translation we represent each node as a NuSMV module. In the module, we declare an enumerated variable, named transition, whose possible values are the list of possible activation functions. Thus, if a node's possible activation functions were $f_0 \cdots f_i$, we would declare:

$$\texttt{transition} : \texttt{f}_0, \texttt{f}_1, ..., \texttt{f}_i;$$

Every node has other nodes it interacts with, and nodes that interact with it. In other words, each node can take the values of some other nodes as inputs into its activation function, and some nodes take its value as an input into their activation functions. From the perspective of the module for the node itself, we only care about the former, the nodes which it must take as inputs into its activation function. For every one of these nodes, whether that interaction be positive or negative, mandatory or optional, a corresponding formal parameter in the module declaration is given. For the special case that a node's value is potentially taken as both an activating and repressing input into the function, that node is only declared once in the formal parameters. For relevant nodes $n_0 \cdots n_i$ of node X, we declare:

$$\texttt{nodeX}(\texttt{n}_0, \texttt{n}_1, \cdots, \texttt{n}_i)$$

For every optional connection that is associated with the node we declare a Boolean variable. For an optional connection with node n, the following variable is declared

$$\texttt{n\_isConnected} : \texttt{boolean};$$

For the special case that a node's value is both activating and repressing and both connections are optional, we will need to define two Boolean variables, $\texttt{n\_isConnectedA}$ and $\texttt{n\_isConnectedR}$. The encoding of the overall effect for activating and repressing connections, unlike the encoding of which connections are optional, is not done explicitly, but rather implicitly in the way in which the connections are factored into the activation functions.

To represent the node's state of activation (active or inactive) at the current time step, we declare:

$$\texttt{value} : \texttt{boolean};$$

Next we specify how the node's value changes from one time step to the next, which will vary based on which of its possible activation functions was chosen to be used. To reflect this, the transition of the $\texttt{value}$ variable is defined using a case statement, with a case defined for each possible value of the transition variable. For possible activation functions $f_0 \cdots f_i$, the transition function for the $\texttt{value}$ variable would be defined as:

```
next(value) := case
transition = f₀ : ...;
transition = f₁ : ...;
...
transition = fᵢ : ...;
esac;
```

There are eighteen possible activation functions, as defined in the RE:IN tool (see Materials and methods of [35]). They can be defined as Boolean expressions composed of the values of the various connecting nodes. The activation functions view separately activators and repressors, but they do not look at the activators and repressors individually. Rather, they consider if all activators are active, some are active, or none are active. Similarly, the functions look to see if all repressors are active, some are, or none are. Additionally, we must deal with what happens when there are no activators that exist at all, or no repressors that exist at all, as non-existence of inputs is dealt with differently than existence with no active inputs. Thus, with four possible states of the activator inputs and four of the repressor inputs, we have sixteen different possible inputs conditions for which the eighteen activation functions must be defined, as reflected in the RE:IN definitions [35].

Since activators and repressors are never viewed individually but rather as an aggregate, that is how they are represented in the Boolean logical expression as well. Accordingly, the activation functions can be expressed purely as combinations of four expressions: the disjunction of all activators, the conjunction of all activators, the disjunction of all repressors, and the conjunction of all repressors. These conditions can give us all the information we need to know while defining the activation functions. To test if all activators are on, we check if their conjunction holds, to test if some are on, we check if their disjunction holds, and to test if all are inactive, we check if the negation of the disjunction holds. Similar considerations are applied to repressors. Using this methodology, we have constructed Boolean expressions that meet each of the eighteen functions' specifications for all the sixteen cases.

The definitions are complicated, however, by the existence of optional connections. The issue this introduces is that if the optional connection is turned off (not selected), it should be removed from the function. A naive work-around is to simply use branching and define a different function when the connection is on versus when it is off, but this can become impractical as the number of optional connections grows. Thus, we use an alternative solution by observing that activators and repressors, as explained above, are only found in lists of conjunctions or disjunctions. Using this observation, we can define logical expressions to accomplish our goal. Let us consider the following example. Assume we have three connecting nodes which serve as activators, named n_0, n_1, and n_2. Further, assume n_0 is optional and n_1 and n_2 are mandatory. Thus, as described above, there exists in the module a Boolean variable n_0_isConnected. The n_0, n_1, and n_2 variables will never appear in isolation, but only as sets in conjunction and disjunction with each other. We will deal with cases of conjunction and disjunction separately. The conjunction expression looks as follows:

$$n_0 \wedge n_1 \wedge n_2$$

We desire the following property: that the value of n_0 only be considered when n_0_isConnected is true, and that otherwise the value of n_0 is ignored. It can be observed that for any Boolean variable X, $X \wedge \text{true} = X$, therefore if we could have a variable N_0, such that when $n_0\_isConnected = true$, $N_0 = n_0$, and when $n_0\_isConnected = false$, $N_0 = true$, we could write:

$$N_0 \wedge n_1 \wedge n_2$$

This statement would solve our issue. When $n_0\_isConnected = true$, N_0 can be replaced with n_0 and we yield $n_0 \wedge n_1 \wedge n_2$, and when $n_0\_isConnected = false$, N_0 can be replaced by true, and we get: $true \wedge n_1 \wedge n_2 = n_1 \wedge n_2$. This variable N_0 can be represented as:

$$n_0 \vee \neg n_0\_isConnected$$

This can be understood by the following property of disjunction: given a variable X, $(X \vee true = true), (X \vee false = X)$. Thus, when $n_0\_isConnected$ is true, we obtain $n_0 \vee false = n_0$, and when $n_0\_isConnected$ is false, we obtain $n_0 \vee true = true$, exactly the property we wanted for N_0. We can now write our conjunction as:

$$(n_0 \vee \neg n_0\_isConnected) \wedge n_1 \wedge n_2$$

As can be readily seen, this expression would require no branching, and would grow linearly as more optional connections are incorporated. We can deal with the case of disjunction in a similar fashion. Consider the expression:

$$n_0 \vee n_1 \vee n_2$$

Based on the observation that $X \vee false = X$, we desire a variable N_0, such that when $n_0\_isConnected$ is true, $N_0 = n_0$, and when $n_0\_isConnected$ is false, $N_0 = false$. This is satisfied by the expression:

$$n_0 \wedge n_0\_isConnected$$

This is based on the observation that for any Boolean variable X, $(X \wedge true = X)$, $(X \wedge false = false)$. The expression in its entirety would be written as:

$$(n_0 \wedge n_0\_isConnected) \vee n_1 \vee n_2$$

Thus, to generalize this, for a set of Boolean variables of which $m_0 \cdots m_i$ are mandatory, and $o_0 \cdots o_j$ are optional based on the values of corresponding variables $o_0\_isConnected \cdots o_j\_isConnected$, we can express their conjunction as:

$$m_0 \wedge \ldots \wedge m_i \wedge (o_0 \vee \neg o_0\_isConnected) \wedge \ldots \wedge (o_j \vee \neg o_j\_isConnected)$$

And their disjunction as:

$$m_0 \vee \ldots \vee m_i \vee (o_0 \wedge o_0\_isConnected) \vee \ldots \vee (o_j \wedge o_j\_isConnected)$$

This approach still contains one additional problem: it fails to capture the required semantics when all activators are optional or when all repressors are optional. In both cases of conjunction and disjunction, we are relying on the optional connections being subsumed into the larger equation. But when all connections are optional, in the case where all of them are turned off there is no larger equation to subsume them. Take the example above, but assume n_0, n_1 and n_2, are all optional. We would then write their disjunction as:

$$(n_0 \wedge n_0\_\texttt{isConnected}) \vee (n_1 \wedge n_1\_\texttt{isConnected}) \vee (n_2 \wedge n_2\_\texttt{isConnected})$$

When all three connections are turned off, this reduces to false. As noted above, nodes with no activators and/or no repressors have different meanings based on which activation functions they are using, and it is incorrect for this expression to evaluate always to false. This issue is also relevant in the conjunction:

$$(n_0 \vee \neg n_0\_\texttt{isConnected}) \wedge (n_1 \vee \neg n_1\_\texttt{isConnected}) \wedge (n_2 \vee \neg n_2\_\texttt{isConnected})$$

The expression would evaluate to true when all connections are turned off, posing the same issue. This problem can be solved by using a combination of both methods, branching and Boolean logic encoding. We can divide all nodes into three classes. There are those that have some mandatory activators (or no activators at all) and some mandatory repressors (or no repressors at all). Their functions can be implemented exactly as described above without any branching.

Then there are functions which have some mandatory repressors (or no repressors at all) but only optional activators, or vice versa. Without loss of generality, let us deal with the case where there are only optional activators and some mandatory repressors. Two functions are given, one the version described above where Boolean logic is used to encode the possible activator connections, one where it is presumed that no activators exist (however that may be defined for that particular function). The branching condition is the disjunction of all connection variables. Continuing the above example, it would be:

$$(n_0\_\texttt{isConnected} \vee n_1\_\texttt{isConnected} \vee n_2\_\texttt{isConnected})$$

When this condition is true, we use the version where we use Boolean logic to ignore connections, as since, based on the condition, at least one connection is true, it will subsume the other connections that aren't. If the condition is false, that means all connections are turned off, and we can use the version that assumes there are no activators.

The third class contains only optional activators and only optional repressors. We can deal with this by branching twice, for a total of four functions.

Using this technique, with a maximum of two branches, we can encode a function that calculates the node's value based on the values of the connection variables in an efficient manner.

Thus, in general: for a node whose activation function contains only optional activators and not only optional repressors (whether that be some mandatory repressors or no repressors at all), or only optional repressors and not only optional activators, if C is the disjunction of all the $o_i\_\texttt{isConnected}$ variables (corresponding to optional interactions), and f_0 is the activation function as described above using Boolean logic to remove optional connections that are inactive, and f_1 is the activation function written as if assuming all optional connections are turned off, we can write the complete activation function as:

$$C \ ? \ f_0 : f_1;$$

For a node containing only optional activators and only optional repressors, if A is the disjunction of all activators and R the disjunction of all repressors, and f_0 is the activation function with Boolean algebra to remove non-active connections as specified above, and f_1 uses Boolean algebra for the activators but assumes there are no repressors, and f_2 uses Boolean algebra for the repressors but assumes there are no activators, and f_3 assumes there are no activators or repressors, we can write the complete function as:

$$A \ ? \ (R \ ? \ f_0 : f_1) : (R \ ? \ f_2 : f_3)$$

Other factors that the node module must consider is whether the node it represents is knocked out (KO) or overexpressed (FE) in a particular experiment. Currently in RE:IN overexpression and knockout properties hold for the duration of the experiment, although in the future this could be extended to model e.g. temperature sensitive knock outs. For every node capable of being overexpressed (as defined in the network file), we define a Boolean variable FE. Then, in the case statement in the value transition, we append this to the top:

FE : TRUE;

Similarly, for nodes capable of being knocked out, we define a Boolean variable KO. Then, in the case statement in the value transition, we append this to the top:

KO : FALSE;

The KO and FE variables transition relation is specified to maintain the current state, thus if KO is true at the beginning of the experiment it will remain true throughout the experiment, while if it is false at the beginning of the experiment it will always remain false, and the same holds for FE.

After creating modules that define all the nodes, we must actually instantiate them. This is done in the main module. When each node is instantiated, the proper values for its formal parameters (which represent the nodes connecting to it) are passed in. The complete structure of a node module is as follows (assuming that the model specification allowed for this node to be knocked out (KO) or over-expressed (FE)):

```
MODULE < name > (< list of node inputs >)
VARS
value : boolean;
transition : < list of possible activation functions >;
KO : boolean;
FE : boolean;
< List of boolean variables corresponding to optional connections >
```

```
ASSIGN
next(value) := case
KO : FALSE;
FE : TRUE;
< All possible values of transition >
```

In NuSMV modules can be composed in a synchronous or asynchronous composition, which allows us to directly support synchronous semantics of Boolean networks, in which all nodes values are updated at each step, or asynchronous semantics for Boolean networks, in which only a single node updates its value at each step, and other nodes values remain unchanged. RE:IN supports both modes and the user needs to specify for each system whether to use synchronous or asynchronous semantics. Research on the pluripotency network and stem cell reprogramming has used synchronous semantics [10,12], while for other systems it has been shown how asynchronous semantics can be gainfully modeled in RE:IN [35].

3   Experiment Encoding in Temporal Logic and Synthesis

We next define how the experimental observations constraints in the RE:IN observations language are translated into temporal logic. In RE:IN the observations are specified in a separate model file (with a .obs extension). The observations consist of experiments, which in turn are divided into time-steps. At each time-step within each experiment, a constraint upon the values of the model at that time (including knocked out or over-expressed nodes) may be specified. A solution for the specification is a concrete network (with set activation functions and connections) that can satisfy all experiments. An experiment is satisfied if there exists a specific initial set of values for the nodes in the network that can lead to the satisfaction of all constraints at all time points. A given concrete network need not have the same initial values satisfy all experiments. Each experiment can be satisfied by a different set of initial values, as long as all experiments are satisfied.

NuSMV allows for specification in both Linear Temporal Logic (LTL) [31] as well as Computational Tree Logic (CTL) [7,13]. Both logics have advantages and disadvantages in terms of expressiveness and algorithmic performance of the corresponding model checking algorithms. LTL only allows specifications over linear paths, whereas CTL allows for specifications over branching behavior, with existential and universal quantification over paths. Intuitively, this makes CTL a natural choice for the synthesis problem we consider here. Since experiments need not have the same initial values, we want to consider all possible sets of initial values, and check if for each experiment, a set exists that satisfies the experiment. This involves quantifying over different possible initial values each node can take on (i.e. different branches of execution). CTL can handle branches in the system dynamics, making it well suited to express solutions to different experiments as different possible branches of initial values.

Unlike the initial values, the activation function and optional connection variables must stay the same for all experiments, so as to have a single concrete model satisfying all experiments. We declare these variables in the model as FROZENVARs, not allowing them to change after the initial state. However, we add a new condition to the transition of the value variables in each node. If the current state is the initial state, we allow the value variable to take on an arbitrary value in the next time state. This is done by creating a variable in the main module named initial that is only true in the initial state, and then appending to the beginning of the case statement in the value transition of each node module:

$$\text{initial} : \{\text{TRUE}, \text{FALSE}\};$$

3.1 CTL Encoding

The CTL expression then asserts that from the initial state of connections and transition functions, there cannot be found for each experiment a branch of initial node values that satisfies the given experiment. If the property does not hold, the counter example given, is a solution to our synthesis problem. In RE:IN it is possible to ask the tool to enumerate all solutions up to a specified limit on the number of solutions. We also support this in NuSMV by disallowing the solution and running NuSMV again. This is repeated until all solutions are found, or the solution limit is reached.

An individual experiment can be encoded as follows: Assuming $c_0 \cdots c_n$ are the constraints placed on the network at time-steps $0 \cdots n$, we can use the CTL EX (exists next) operator and write:

$$\text{EX}(c_0 \wedge (\text{EX}c_1(... \wedge (\text{EX}c_n))))$$

A CTL property that specifies that there is a solution for all experiments $e_0 \cdots e_i$, assuming each experiment is encoded as above, can be formulated as follows:

$$(e_0 \wedge e_1 \wedge ... \wedge e_i)$$

We can then assert in NuSMV the negation of the formula and use the CTL model checking algorithms to check it:

$$\text{CTLSPEC} \neg (e_0 \wedge e_1 \wedge ... \wedge e_i)$$

If the property does not hold, the CTL model checking algorithm finds a counter example which corresponds to a concrete Boolean network satisfying all experimental observations. Otherwise if the property holds, the CTL model checker proves this property and the conclusion is that no consistent Boolean networks exist.

We have successfully implemented this translation in our prototype tool, and identified that the CTL approach had two initial drawbacks we encountered during evaluation. First, since NuSMV only supports verifying CTL specifications with Binary Decision Diagrams (BDDs), and BDDs are very sensitive to variable order and can grow anywhere from linearly to exponentially in size depending upon it, additional work will be needed to identify optimal variable ordering and reordering heuristics that may work well for our network synthesis problem. For a toy model available with RE:IN and for the myeloid progenitor differentiation [35], our prototype implementation can solve the synthesis problem in a few seconds on a standard laptop. However it should be noted that the performance is still slower than the current RE:IN solvers based on the Z3 SMT solver. For very large problems the BDD often got too large to be feasible, specifically for the pluripotency models studied in [12]. It still remains open whether improved variable orderings and use of heuristics and alternative CTL model checking algorithms can enable to scale this approach to handle more complex problems.

The second drawback is that NuSMV counter examples currently only provide partial information, i.e. they only provide a single path of execution. Thus, although the counter example gives us the most crucial details of the solution, such as which nodes were connected, and which transition functions were used (to form a concrete network), we cannot reconstruct from the counter example what initial values were found to be satisfactory for each experiment, as these exist in different possible branches the model could have taken. We have not yet addressed this issue in our current implementation, it could be either tackled by trying to obtain more information in the counter example, or by setting the inferred activation functions and optional interactions that give a consistent solution and invoking additional model checking queries.

3.2 LTL Encoding

This leads us to Linear Temporal Logic (LTL). At first glance, LTL does not seem to be a natural way to specify the required behavioral properties. LTL can only quantify over a linear path, whereas we need to quantify over different branches, as a linear path would not allow different experiments to have differing initial values. We suggest to solve this problem by viewing the experiments happening not in parallel as different executions, but sequentially. For example, let's say we have two experiments, both lasting 20 time steps. Instead of assuming that time-steps 1–20 (0 is reserved for setting connection and function values) are when both experiments occur and exploring different branches with different initial node values as we did above, we assume that the first experiment takes place in time-steps 1–20, and the second one in time-steps 21–40. To allow this, we must let our `value` variables take on an arbitrary value at time-step 21. To accomplish this we replace our Boolean variable `initial` with an integer named `counter` that counts what number time-step we are at. Then, in the value transition case statement, we append to the top the following:

$$(\texttt{counter} = 0) \lor (\texttt{counter} = 20) : \{\text{TRUE}, \text{FALSE}\};$$

Since this is defining what `value` will be in the next state, at time-steps 1 and 21 `value` can now be set nondeterministically to true or false.

In general, for a set of experiments $e_0 \cdots e_i$, when for any experiment e_x, t_x is the duration of e_x, we define a set of time-steps, $s_0 \cdots s_i$, where $s_0 = 0$, and

$s_n = s_{n-1} + t_{n-1}$. Then we append the following to the beginning of the value transition case statement:

$$(\texttt{counter} = \texttt{s}_0) \lor (\texttt{counter} = \texttt{s}_1) \lor ... \lor (\texttt{counter} = \texttt{s}_i) : \{\texttt{TRUE}, \texttt{FALSE}\};$$

Now that the experiments are not in parallel, but rather take place in a single run, LTL properties can be used to specify the desired requirement that for the initial transitions and connections chosen, there are initial node values for each experiment that satisfy the observations. An individual experiment can be encoded as follows: Assuming $c_0 \cdots c_n$ are the constraints placed on the network at time-steps $0 \cdots i$, we can use the LTL X (next operator) and write (note the parenthesis are left open and should be closed by composing the formula with the formula of the next experiment):

$$\texttt{X}(c_0 \land (\texttt{X}c_1(... \land (\texttt{X}c_n$$

To find a solution for all experiments $e_0 \cdots e_i$, assuming each experiment is encoded as above, we assert:

$$\texttt{LTLSPEC}\neg(e_0 \land e_1 \land ... \land e_i$$

Followed by the appropriate amount of closing parenthesis to balance the opening ones. This mode of running our prototype implementation was called LTL Run, or LTLR. For LTL model checking NuSMV supports both a method using BDDs and a bounded model checking approach based on SAT solving. The two challenges with CTL discussed earlier are strong points here. Bounded model checking of LTL specifications based on SAT does not need to deal with sensitivity to variable ordering of BDDs, and on larger problems this approach seemed to yield a significant speedup when compared to BDDs. It should be noted that a more careful comparison should be done after evaluating the performance gained by automatic optimizations to variable ordering and reordering. Additionally, since there is no branching, the counterexample trace gives us full information, including the different initial node values that satisfied each experiment and the full trajectory.

In our evaluation, the LTLR method of specification gives optimal performance when the Abstract Boolean Network was relatively large, but the experiment specifications total length was not too long. However, when there are many experiments being tested, the bounds on the bounded model checking rises linearly. For example, if there are 20 experiments whose duration is 20 timesteps, the bounds will be 400. Large values for the bound in the bounded model checking significantly effected the performance and make synthesis challenging for systems with larger number of experiments.

This led to a new idea, termed LTL parallel, or LTLP. Instead of stacking the experiments into a linear trajectory, we have them run in parallel. Unlike in CTL where the model conceptually only evaluates one experiment at a time, but we can check them all in parallel using branching, here we check them all in parallel without branching to different possible initial values by introducing redundant value variables in each node, one per experiment. Each experiment is verified using its own corresponding value variables, according to the transition function and connection values chosen in the initial state. Thus, assuming we have experiments $0 \cdots i$, we declare for each node a set of value variables $v_0 \cdots v_i$, and define for each one the same transition relation, as described above. Now each experiment can be evaluated with different initial values and different trajectories.

An individual experiment can be encoded as follows: For experiment e_i, let $c_0 \cdots c_n$ be the constraints placed on the values of the v_i variables of the network at time-steps $0 \cdots n$. We can use the LTL X (next operator) and write:

$$X(c_0 \wedge (Xc_1(\dots \wedge (X \wedge c_n))))$$

The entire specification would then be, for experiments $e_0 \cdots e_i$:

$$\text{LTLSPEC} \neg (e_0 \wedge e_1 \wedge \dots \wedge e_i)$$

While this reduces the bounds on bounded model checking to the size of the largest experiment duration, it increases the complexity of the model. This approach performed more favorably than the LTLR mode on models with many experiments. In particular, for some of the challenging synthesis problems involving the embryonic stem cell gene network [10,12] we obtained solutions in only a few minutes, is some cases outperforming the reference Z3 based synthesis implementation of RE:IN. While these are only preliminary results and our algorithm's performance need to be evaluated systematically and rigorously on a wide range of benchmarks, these initial results are encouraging and can in the long term yield complementary algorithms and approaches to tackle the inherent complexity involved in synthesizing experimentally constrained gene regulatory networks.

4   Related Work

Formal reasoning methods enable to mathematically prove the correctness of a model with respect to a specification. In the biological context, formal reasoning is complementary to Bayesian inference and machine learning approaches [16, 26,27]. Compared to testing, which is mainly based on simulation, verification enables to obtain a proof, while simulation can only increase our confidence in the correctness of an implementation. As a result, and due to the significant research breakthroughs in scalability of algorithmic methods, formal verification is used strategically for important systems in industry [1,15,21]. Temporal logic [7,13,31] has been shown to be especially suitable for specifying and reasoning about reactive software and hardware. There are different variants of temporal logic, including linear temporal logic (LTL) [31] and computational temporal logic (CTL) [7], while CTL* [13] is a temporal logic that is a superset of LTL and CTL.

While temporal logic is now being actively used also in the biological context [3–5,9,14,33], important open questions remain regarding the appropriate type of temporal logic needed, and the complexity of key verification and synthesis [25] algorithms. Developing synthesis methods for biological systems is an active area of research [17,19,20,22,23,29,30,34], synthesis methods can automate the process of model development constrained by experimental data and enable rapid construction of predictive models. The inherent complexity of synthesis methods is a major challenge that needs to be addressed to make synthesis algorithms more broadly applicable in biology.

In recent work the RE:IN tool has been extended to analyze network motifs [24]. In the context of network motifs, in [2] certain motifs and their dynamic properties are characterized using temporal logic and parallel model checking is used to verify properties of networks with around ten components. In [18] approximate methods for analyzing gene regulatory networks are developed utilizing network motifs.

Acknowledgment. The research was partially supported by the Horizon 2020 research and innovation programme for the Bio4Comp project under grant agreement number 732482. We also acknowledge the support of the program for Summer Science Research Internships for Yeshiva University Students at Bar-Ilan University.

References

1. Ball, T., Levin, V., Rajamani, S.K.: A decade of software model checking with slam. Commun. ACM **54**(7), 68–76 (2011)
2. Barnat, J., Brim, L., Cerna, I., Drazan, S., Safranek, D.: From simple regulatory motifs to parallel model checking of complex transcriptional networks. In: Pre-proceedings of Parallel and Distributed Methods in Verification (PDMC 2008) Budapest, pp. 83–96 (2008)
3. Bartocci, E., Lió, P.: Computational modeling, formal analysis, and tools for systems biology. PLoS Comput. Biol. **12**(1), e1004591 (2016)
4. Batt, G., et al.: Validation of qualitative models of genetic regulatory networks by model checking: analysis of the nutritional stress response in *Escherichia coli*. Bioinformatics **21**, 19–28 (2005)
5. Chabrier, N., Fages, F.: Symbolic model checking of biochemical networks. In: Priami, C. (ed.) CMSB 2003. LNCS, vol. 2602, pp. 149–162. Springer, Heidelberg (2003). https://doi.org/10.1007/3-540-36481-1_13
6. Cimatti, A., Clarke, E., Giunchiglia, F., Roveri, M.: NUSMV: a new symbolic model checker. Int. J. Softw. Tools Technol. Transf. **2**(4), 410–425 (2000)
7. Clarke, E.M., Emerson, E.A.: Design and synthesis of synchronization skeletons using branching time temporal logic. In: Kozen, D. (ed.) Logic of Programs 1981. LNCS, vol. 131, pp. 52–71. Springer, Heidelberg (1982). https://doi.org/10.1007/BFb0025774
8. de Moura, L., Bjørner, N.: Z3: an efficient SMT solver. In: Ramakrishnan, C.R., Rehof, J. (eds.) TACAS 2008. LNCS, vol. 4963, pp. 337–340. Springer, Heidelberg (2008). https://doi.org/10.1007/978-3-540-78800-3_24
9. Dubrova, E., Teslenko, M., Ming, L.: Finding attractors in synchronous multiple-valued networks using SAT-based bounded model checking. In: 40th IEEE International Symposium on Multiple-Valued Logic (ISMVL), pp. 144–149 (2010)

10. Dunn, S.-J., Li, M.A., Carbognin, E., Smith, A.G., Martello, G.: A common molecular logic determines embryonic stem cell self-renewal and reprogramming. bioRxiv, p. 200501 (2017)
11. Dunn, S.-J., Li, M.A., Carbognin, E., Smith, A.G., Martello, G.: A common molecular logic determines embryonic stem cell self-renewal and reprogramming. EMBO J. **38**, e100003 (2018)
12. Dunn, S.-J., Martello, G., Yordanov, B., Emmott, S., Smith, A.G.: Defining an essential transcription factor program for naïve pluripotency. Science **344**(6188), 1156–1160 (2014)
13. Emerson, E.A., Halpern, J.Y.: 'Sometimes' and 'not never' revisited: on branching time versus linear time. J. ACM **33**, 151–178 (1986)
14. Fisman, D., Kugler, H.: Temporal reasoning on incomplete paths. In: Margaria, T., Steffen, B. (eds.) ISoLA 2018. LNCS, vol. 11245, pp. 28–52. Springer, Cham (2018). https://doi.org/10.1007/978-3-030-03421-4_3
15. Fix, L.: Fifteen years of formal property verification in intel. In: Grumberg, O., Veith, H. (eds.) 25 Years of Model Checking. LNCS, vol. 5000, pp. 139–144. Springer, Heidelberg (2008). https://doi.org/10.1007/978-3-540-69850-0_8
16. Friedman, N., Linial, M., Nachman, I., Pe'er, D.: Using Bayesian networks to analyze expression data. J. Comput. Biol. **3**(7), 601–620 (2000)
17. Guziolowski, C., et al.: Exhaustively characterizing feasible logic models of a signaling network using answer set programming. Bioinformatics **29**(18), 2320–2326 (2013)
18. Ito, S., Ichinose, T., Shimakawa, M., Izumi, N., Hagihara, S., Yonezaki, N.: Formal analysis of gene networks using network motifs. In: Fernández-Chimeno, M., et al. (eds.) BIOSTEC 2013. CCIS, vol. 452, pp. 131–146. Springer, Heidelberg (2014). https://doi.org/10.1007/978-3-662-44485-6_10
19. Koksal, A.S.: Program Synthesis for Systems Biology. PhD thesis, University of California at Berkeley. Technical Report No. UCB/EECS-2018-49 (2018)
20. Koksal, A.S., Pu, Y., Srivastava, S., Bodik, R., Fisher, J., Piterman, N.: Synthesis of biological models from mutation experimentss. In: SIGPLAN-SIGACT Symposium on Principles of Programming Languages. ACM (2013)
21. Kroening, D., Strichman, O.: Decision Procedures, vol. 5. Springer, Heidelberg (2008)
22. Kugler, H., Plock, C., Roberts, A.: Synthesizing biological theories. In: Gopalakrishnan, G., Qadeer, S. (eds.) CAV 2011. LNCS, vol. 6806, pp. 579–584. Springer, Heidelberg (2011). https://doi.org/10.1007/978-3-642-22110-1_46
23. Kugler, H., Pnueli, A., Stern, M.J., Hubbard, E.J.A.: "Don't care" modeling: a logical framework for developing predictive system models. In: Grumberg, O., Huth, M. (eds.) TACAS 2007. LNCS, vol. 4424, pp. 343–357. Springer, Heidelberg (2007). https://doi.org/10.1007/978-3-540-71209-1_27
24. Kugler, H., Dunn, S.-J., Yordanov, B.: Formal analysis of network motifs. In: Češka, M., Šafránek, D. (eds.) CMSB 2018. LNCS, vol. 11095, pp. 111–128. Springer, Cham (2018). https://doi.org/10.1007/978-3-319-99429-1_7
25. Kupferman, O.: Recent challenges and ideas in temporal synthesis. In: Bieliková, M., Friedrich, G., Gottlob, G., Katzenbeisser, S., Turán, G. (eds.) SOFSEM 2012. LNCS, vol. 7147, pp. 88–98. Springer, Heidelberg (2012). https://doi.org/10.1007/978-3-642-27660-6_8
26. Liang, S., Fuhrman, S., Somogyi, R.: Reveal, a general reverse engineering algorithm for inference of genetic network architectures. In: Pacific Symposium on Biocomputing, vol. 3, pp. 18–29. Springer (1998)

27. Marbach, D., Prill, R.J., Schaffter, T., Mattiussi, C., Floreano, D., Stolovitzky, D.: Revealing strengths and weaknesses of methods for gene network inference. Proc. Nat. Acad. Sci. **107**(14), 6286–6291 (2010)
28. Mishra, A., et al.: A protein phosphatase network controls the temporal and spatial dynamics of differentiation commitment in human epidermis. Elife **6**, e27356 (2017)
29. Moignard, V., et al.: Decoding the regulatory network of early blood development from single-cell gene expression measurements. Nat. Biotechnol. **33**(3), 269 (2015)
30. Paoletti, N., Yordanov, B., Hamadi, Y., Wintersteiger, C.M., Kugler, H.: Analyzing and synthesizing genomic logic functions. In: Biere, A., Bloem, R. (eds.) CAV 2014. LNCS, vol. 8559, pp. 343–357. Springer, Cham (2014). https://doi.org/10.1007/978-3-319-08867-9_23
31. Pnueli, A.: The temporal logic of programs. In: Proceedings of 18th IEEE Symposium on Foundations of Computer Science, pp. 46–57 (1977)
32. Shavit, Y., et al.: Automated synthesis and analysis of switching gene regulatory networks. Biosystems **146**, 26–34 (2016)
33. Tiwari, A., Talcott, C., Knapp, M., Lincoln, P., Laderoute, K.: Analyzing pathways using SAT-based approaches. In: Anai, H., Horimoto, K., Kutsia, T. (eds.) AB 2007. LNCS, vol. 4545, pp. 155–169. Springer, Heidelberg (2007). https://doi.org/10.1007/978-3-540-73433-8_12
34. Woodhouse, S., Piterman, N., Wintersteiger, C.M., Göttgens, B., Fisher, J.: SCNS: a graphical tool for reconstructing executable regulatory networks from single-cell genomic data. BMC Syst. Biol. **12**(1), 59 (2018)
35. Yordanov, B., Dunn, S.-J., Kugler, H., Smith, A., Martello, G., Emmott, S.: A method to identify and analyze biological programs through automated reasoning. NPJ Syst. Biol. Appl. **2**, 16010 (2016)
36. Yordanov, B., Wintersteiger, C.M., Hamadi, Y., Kugler, H.: SMT-based analysis of biological computation. In: Brat, G., Rungta, N., Venet, A. (eds.) NFM 2013. LNCS, vol. 7871, pp. 78–92. Springer, Heidelberg (2013). https://doi.org/10.1007/978-3-642-38088-4_6

On the Existence of Synergies and the Separability of Closed Reaction Networks

Tomas Veloz[1,2]([✉]), Alejandro Bassi[1,4], Pedro Maldonado[1], and Pablo Razeto[1,3]

[1] Instituto de Filosofía y Ciencias de la Complejidad, Los Alerces, 3024 Ñuñoa, Chile
tveloz@gmail.com
[2] Departamento Ciencias Biologicas, Facultad Ciencias de la Vida, Universidad Andres Bello, 8370146 Santiago, Chile
[3] Vicerrectoría Académica, Universidad Diego Portales, Manuel Rodríguez Sur 415, Santiago, Chile
[4] Laboratorio de Sueño y Cronobiología, Programa de Fisiología y Biofísica, Instituto de Ciencias Biomédicas, Universidad de Chile, Av. Independencia 1027, Independencia, Santiago, Chile

Abstract. It is well known that closure is a necessary topological property for a reaction network to be dynamically stable. In this work we combine notions of chemical organization theory with structural properties of reaction networks to distill a minimal set of closed reaction networks that encodes the non-trivial stable dynamical regimes of the network. In particular, these non-trivial closed sets are synergetic, in the sense that their dynamics cannot always be computed from the dynamics of its closed constituents. We introduce a notion of separability for reaction networks and prove that it is strictly related to the notion of synergy. In particular, we provide a characterization of the non-trivial closed reaction networks by means of their degree of internal synergy. The less trivial the dynamics of the reaction network, the less can be separated into constituents, and equivalently the more synergies the reaction network has. We also discuss the computational and analytical benefits of this new representation of the dynamical structure of a reaction network.

Keywords: Reaction networks · Chemical organization theory · Self-organization · Closure · Synergy · Separability

1 Introduction

Reaction networks is one of the most important representational languages in systems biology, and depending on the way in which the dynamics is defined (e.g. discrete, stochastic, continuous), it has been proven to be equivalent to other formal languages that have been developed in the context of concurrent information processing such as Vector Addition Systems, Commutative Grammars [20],

© Springer Nature Switzerland AG 2019
M. Chaves and M. A. Martins (Eds.): MLCSB 2018, LNCS 11415, pp. 105–120, 2019.
https://doi.org/10.1007/978-3-030-19432-1_7

Petri Nets [15], and others. A particular perspective that focuses in the relation between the structure, i.e. topological and stoichiometric properties, of the reaction network and its dynamical stability, i.e. the existence of attractors and stable manifolds in the phase space (either for discrete, stochastic, or continuous dynamics), known as Chemical Organization Theory (COT) [5], has been shown to be applied to model a wide range of biochemical systems [1,2,9,10], to clarify the relation between the reaction network's structure and its dynamical evolution [8,13,14,23], to provide novel algorithms for the study of reaction networks [2,3,18], and it has also been proposed as a basis for chemical computing [11,12]. Moreover, the scope of COT does not restrict to biochemical systems. COT has also been applied to model non-biochemical systems such as ecological [24], decision [21] and political systems [4]. The latter has motivated proposing reaction networks as a candidate language to express systems in the general sense of systems theory [22,23].

In COT, chemical organizations are collections of species forming subnetworks that are both closed and self-maintaining. The closure property can be verified in a computationally much more easy manner than self-maintenance. For this reason, algorithms that calculate chemical organizations of a reaction network prefer to verify first closure and then self-maintainance [2]. However, the closure property has been poorly studied despite some basic promising results that show that the set of closed sets forms a lattice with respect to some suitable join and meet operators [5,18]. In this work we will retake the mathematical study of closed sets and will show that there is a minimal set of closed sets that encodes all the non-trivial aspects of the dynamics of the reaction network. Such non-triviality corresponds from a structural point of view to synergies within the reaction network. A synergy occurs when the combination of two reaction networks is able to trigger novel reactions, that is reactions that cannot be triggered by any of the reaction networks being combined. Note that the aim of this article is not in the development of new algorithms, but in the preparation of a theoretical ground for understanding what structural properties can be of use for novel and more efficient algorithms.

The paper is organized as follows: In Sect. 2 we introduce the basic elements of COT with a focus on the closure property. In Sect. 3 we show different ways in which a closed set is not relevant from a dynamical point of view (and thus should not be computed by an algorithm that aims at understanding the dynamical structure of a reaction network). In Sect. 4 we focus on the structure of the relevant (synergetic) reaction networks and study the relation between synergy and a novel notion of non-separability, which represents the extreme case of complete synergy within the reaction network. In Sect. 5 we summarize and discuss our results and propose further lines of research as well as possible applications.

2   Closed Reaction Networks

2.1   Reaction Networks: Static Concepts

Let $(\mathcal{M}, \mathcal{R})$ be a reaction network with $\mathcal{M}$ a set of species and $\mathcal{R}$ a set of reactions.

For each $r \in \mathcal{R}$, let supp(r) to be the set of reactants of r, and let prod(r) the set of products of r.

Definition 1. *Let $X \subseteq \mathcal{M}$. We define $\mathcal{R}_X \subseteq \mathcal{R}$ as the set of reactions $r \in \mathcal{R}$ such that supp(r) $\subseteq X$. $\mathcal{R}_X$ is named the set of reactions associated to X, $(X, \mathcal{R}_X)$ is a sub-network of $(\mathcal{M}, \mathcal{R})$.*

In an aggregated manner we define supp($\mathcal{R}_X$) $= \cup_{r \in \mathcal{R}_X}$ supp(r), prod($\mathcal{R}_X$) $= \cup_{r \in \mathcal{R}_X}$ prod(r).

Similarly, consider a collection of sets $\boldsymbol{X} = \{X_1, ..., X_k\}$. We define $\mathcal{R}_{\boldsymbol{X}} \subseteq \mathcal{R}$ as the set of reactions $r \in \mathcal{R}$ such that supp(r) $\subseteq \bigcup_{X \in \boldsymbol{X}} X$, and supp($\mathcal{R}_{\boldsymbol{X}}$) $= \cup_{r \in \mathcal{R}_{\boldsymbol{X}}}$ supp(r), prod($\mathcal{R}_{\boldsymbol{X}}$) $= \cup_{r \in \mathcal{R}_{\boldsymbol{X}}}$ prod(r).

The previous definition allows to consider a reaction network either from a set of species or from a collection of sets of species.

Definition 2. *Let $X \subseteq \mathcal{M}$. A species $s \in \mathcal{M}$ is reactive with respect to X if and only if there exists $r \in \mathcal{R}_{X \cup \{s\}}$ such that $s \in$ supp(r) $\cup$ prod(r). X is reactive if and only if every species $s \in X$ is reactive with respect to X.*

Reactive reaction networks are those where every species participate either as reactant or as product in at least one reaction.

Definition 3. *X is closed if and only if prod($\mathcal{R}_X$) $\subseteq X$.*

The defining feature of closed sets is that there is no qualitative novelty in their dynamics. Indeed if X is not closed, the dynamical operation of X is going to add species that are in prod($\mathcal{R}_X$) but not in X. This in turn will activate new reactions in the system[1], whose products might successively add new species until a closed set is reached. Because we are dealing with a finite $\mathcal{M}$ this is always the case. In realistic dynamics it is possible that some species disappear over time. However, in any case we can ensure that in the long run the set of reactive species will form either a closed set that contains X or a closed set that is contained by X [5]. Therefore, closure is a necessary condition for a set of species to have a stable dynamics such as fixed point, periodic orbit or limit cycle [14].

Definition 4. *The set of closed sets $\mathcal{C}$ of a reaction network $(\mathcal{M}, \mathcal{R})$ is called the closure structure of $(\mathcal{M}, \mathcal{R})$.*

2.2 Reaction Networks: Operational Considerations

Definition 5. *Let $G_{CL}(X)$ be the closed set of smallest cardinality containing X. We call $G_{CL}(X)$ the generated closure of X.*

---

[1] Those in $\mathcal{R}_{X \cup \text{prod}(\mathcal{R}_X)}$.

Lemma 1. $G_{CL}(X)$ *is unique.*

Proof. Since $\mathcal{M}$ is finite, we have that $G_{CL}(X)$ can be trivially obtained by successively adding the products of the reactions that are novel. Indeed, let $\mathrm{prod}_*(X) = X \cup \mathrm{prod}(X)$. Since $\mathcal{M}$ is finite, we have that $G_{CL}(X) \subseteq \mathcal{M}$. Hence $G_{CL}(X) = X \cup \mathrm{prod}(X) \cup \mathrm{prod}(X \cup \mathrm{prod}(X)) \cup \cdots = \mathrm{prod}_*^l(X)$, for some $l \in \mathbb{N}$. $\qquad\square$

The generated closure allows to identify, from any set X, the minimal closed set that contains X. In addition, closed sets can be combined to form new closed sets. In particular, for the union and intersection of any two closed sets a new closed set can be generated using G_{CL}.

Definition 6. *Let* $X \vee_{CL} Y = G_{CL}(X \cup Y)$ *and* $X \wedge_{CL} Y = G_{CL}(X \cap Y)$.

Note that in general $X \vee_{CL} Y \neq X \cup Y$ (and $X \wedge_{CL} Y \neq X \cap Y$). These operators exhibit an interesting feature. Namely, the closure structure $\mathcal{C}$ is in direct correspondence with an order theoretical structure known as *lattice* [6].

Lemma 2. $(\mathcal{C}, \vee_{CL}, \wedge_{CL})$ *is a bounded lattice.*

Proof. See [5], p. 1204. $\qquad\square$

Lemma 2 shows that the closure structure can be equipped with a mathematical structure that allows to safely combine closed sets (using the proper join and meet operations) to build other closed sets. In particular, this fact has been used in [18] to develop an algorithm that can be employed to compute the closure structure[2].

2.3   Reaction Networks: Dynamical Concepts

With the concepts introduced at this point, we are able to identify the closed sets of a reaction network, but we cannot establish a connection between closed sets and the dynamics of the reaction network. In order to model the dynamics of a reaction network, we must equip the reaction network with a way to calculate the occurrence of the reactions within the reaction network. For this reason we introduce the abstract notion of *process* [22]. A process specifies which reactions (and how many or at which rate if needed) occur within a certain time interval, and thus a set of possible processes characterizes what are the *collective trans-formation of species* that can possibly happen in the reaction network. Hence, a process is an abstraction of a particular dynamics, and the set of possible processes is an abstraction of a kynetic law. The representation of a process varies depending of the focus on the study of the reaction network. If our focus is at a purely structural (topological) level, for example if we are interested in properties such as connectivity or centrality of the species, a process is sufficiently

---

[2] The algorithm presented in [18] is employed to build a restricted form of closed sets, called organizations, for specific class of reaction networks, known as flow systems, in which organizations have a lattice structure.

well-specified by a set of reactions, whereas if our focus is on understanding a one-step discrete dynamics or the stoichiometric aspects of the reaction network, a process is sufficiently specified by a vector of (positive) real coefficients. In addition, if our focus is on the time evolution of a reaction network, a process needs to be specified by a vector of functions determined by a particular kynetic law.

The fact that we have different ways to specify a process is crucial from an algorithmic point of view, because, on the one hand, the structural, stoichiometric, and dynamical levels comprise increasingly complex computational problems, and on the other hand, one can make use of the information gathered at a less complex algorithmic level (e.g. topological level) to facilitate the analysis at a more complex algorithmic level (e.g. dynamical level). The aim of COT is to exploit the relations between the less complex and more complex levels to obtain a computationally feasible way to study the dynamical properties of large reaction networks where not only analytic results but also computational simulation are hard to perform. For a detailed explanation of the relations between topological, stiochiometric and dynamical levels, as well as for an introduction to COT see [5, 22].

Most of COT advances have exploited the relation between the stoichiometric and dynamical levels. A fundamental concept to understand such relation is the stoichiometric matrix. The stoichiometric matrix $\mathbf{S}$ encodes the way in which species are consumed and produced by the reactions in the of a reaction network $(\mathcal{M}, \mathcal{R})$. Given a set $X \subseteq \mathcal{M}$ we denote the reduced stoichiometric matrix of the subnetwork $(X, \mathcal{R}_X)$, which only considers the reactions in $\mathcal{R}_X$ by $\mathbf{S}_X$.

Since the stoichiometric matrix encodes the total amount of produced and consumed species involved by the reactions, we will represent a process $\mathbf{v}$ by a non-negative valued real vector in which its i-th coordinate $\mathbf{v}[i]$ specifies either the number of times, or the probability, or the rate at which $r_i \in \mathcal{R}$ occurs, depending if we consider discrete, stochastic or continuous dynamics.

For example in discrete dynamics, let the state of a reaction network by a vector $\mathbf{x}$ of non-negative coordinates where $\mathbf{x}_t[j]$ corresponds to the number of species of type s_j in the reaction network, $j = 1, ..., m$, at some time-step τ. We have that the state $\mathbf{x}_{\tau+1}(\mathbf{v})$ of the reaction network associated to a state $\mathbf{x}$ and a process $\mathbf{v}$ occurring between the time-step τ and $\tau + 1$ is given by the following equation:

$$\mathbf{x}_{\tau+1}(\mathbf{v}) = \mathbf{x}_\tau + \mathbf{S}\mathbf{v}. \tag{1}$$

Equation (1) provides a formal description for the change of the number of species driven by a process $\mathbf{v}$ [7].

Definition 7. *Let v be a non-null process. X is weak-self-maintaining with respect to v if and only if $x_v[j] \geq x[j]$, $j = 1, ..., m$. If, additionally, such process satisfies $v[i] > 0$ if and only if $r_i \in \mathcal{R}_X$, we say X self-maintaining.*

For a weak-self-maintaining set X, there are processes that lead to a non-negative production of all the species involved in the process. These processes, however, might not execute all the reactions in $\mathcal{R}_X$. A well known type of weak-self-maintaining processes are known as elementary modes [16]. Indeed, there is a whole area of study of weak-self-maintaining and other types of processes known as metabolic pathway analysis [17]. Without digging much in the study of metabolic pathways, we would like to notice a very important fact: For self-maintaining sets we are able to find processes such that every reaction in $\mathcal{R}_X$ occurs, and the result of the process does not lead to the consumption of any species. Therefore, self-maintaining sets entail the parts of the reaction network where realistic[3] self-sustainable processes, at a quantitative level of description, can occur.

For continuous dynamics, the state vector is a function of time $\mathbf{x}(t) = (x_1(t), ..., x_m(t))$, where $x_j(t)$ encodes the number of species s_j at time t. By setting the difference between τ and $\tau + 1$ infinitely small, and making $\mathbf{v} \equiv \mathbf{v}(\mathbf{x})$ dependent on the species concentration, Eq. (1) becomes the differential equation

$$\dot{\mathbf{x}} = \mathbf{S}\mathbf{v}(\mathbf{x}), \tag{2}$$

with initial conditions specified by $\mathbf{x}(t_0)$, and it is called a reaction system.

Chemical Organization Theory [5] introduced the crucial notion of organization which connects the stoichiometric analysis with the properties of the reaction system.

Definition 8. *X is an organization if and only if X is closed and self-maintaining.*

An organization satisfies simultaneously the structural property of closure and the stoichiometric-level property of self-maintaining. These two requirements provide a necessary condition for stable dynamics.

Definition 9. *Let $\mathcal{P}(\mathcal{M})$ be the power set of $\mathcal{M}$ and*

$$\phi(t) : \mathbb{R}_{\geq 0}^m \to \mathcal{P}(\mathcal{M}), \quad \mathbf{x}(t) \mapsto \phi(\mathbf{x}(t)) \equiv \{s_i \in \mathcal{M} : x_i(t) > 0\}. \tag{3}$$

*For a state $\mathbf{x}(t) \in \mathbb{R}_{\geq 0}^m$, the set $\phi(\mathbf{x}(t))$ is the **abstraction** of $\mathbf{x}(t)$. For a given set of species $X \subseteq \mathcal{M}$, a state $\mathbf{x}(t) \in \mathbb{R}_{\geq 0}^m$ is an **instance** of X if and only if its abstraction equals X.*

The notions of abstraction and instance connect the representations of the reaction network with the reaction system, and organizations represent the abstractions of all the possible stable instances:

Theorem 1. *If $\mathbf{x}$ is a fixed-point of the ODE (2), i.e., $\mathbf{S}\mathbf{v}(\mathbf{x}) = \mathbf{0}$, then the abstraction $\phi(\mathbf{x})$ is an organization [5].*

Fixed points are the simplest stable behavior of a dynamical system. Identifying fixed points is a fundamental task to understand the dynamics of a system [19]. Thus, Theorem 1 provides a necessary condition for a set of species to form a sub-network with a stable behavior. Indeed, in [14], Theorem 1 is

---

[3] In the sense that none of the reactions in the network is assigned with a zero value in the process.

extended to most stable behaviors, including periodic orbits and limit cycles. In addition, the feasibility of self-maintaining flux vectors is explored in [13], the dynamics of a system with a small number of particles is studied in [9], and the analysis of several metabolic and other biochemical systems can be found in [1,8,10].

2.4   Reaction Networks and Organizations: The Algorithmic Problem

Theorem 1 implies that the identification of organizations of a reaction network is a powerful task to understand its stable dynamics. Moreover, Lemma 2 identifies an elegant way to represent the closure structure. Although this representation has been the basis for the algorithmic developments to compute the closure and the set of organizations [2,18], from an algorithmic point of view computing organizations is not considered to be solved because the potential candidates to be a closed set (and thus a organization) for a reaction network with n species is 2^n. However, when reaction networks in nature have been tested, it has been found that the number of organizations is very small compared to 2^n [2,5]. Therefore, understanding the inner structure of the closed reaction networks is an important task to develop efficient algorithms to compute the set of organizations. While few structural analysis for organizations exist in the literature [23], no article up to date has analyzed the inner structure of closed sets and its relation with the computation of organizations. In this article we will focus on the rather 'simpler' problem of identifying closed sets instead of organizations, and will elucidate that most of the combinatorial problems that seem to be inherent to the computation of organizations can be avoided by taking a closer look to the inner structure of closed sets. In particular, we will note that there are various ways in which a closed set is dynamically irrelevant, and thus we can *a priori* discard those sets as candidates to be organizations. By a closed set X being dynamically irrelevant we mean that the set can be represented by either one or many closed sets that are contained in X such that the dynamics of X can be directly computed from the dynamics of those sets contained in X. Remarkably, the founders of COT were aware of and made explicit that structural analysis of COT fundamental properties would lead to the development of efficient algorithms in their seminal paper [5], but no analysis of the structure of closed sets has been done up to now. We would like to clarify that in this work we will not focus on the development of algorithms, but on identifying relevant structural results that can be later applied to develop efficient algorithms.

3   When a Closed Set Is Dynamically Relevant?

In this section we will show that the notion of closure is *too broad to be a topological criteria for the stability of a reaction network*, and thus captures types of sets whose reaction networks are not relevant from a dynamical point of view. For simplicity, we will introduce three types of closed sets whose dynamic irrelevance can be made clear by increasingly less trivial reasons, and thus we will be able to formally introduce the notion of dynamic irrelevance for a closed set. Identifying the dynamically relevant closed sets on the one hand allows to clarify the inner structure of closed sets and its relation with the dynamics of the reaction network, and on the other hand can be applied to develop more efficient algorithms to compute the closure structure.

3.1   Non-reactive Closed Sets Are Irrelevant

Let X be closed and s not reactive with respect to X, then $X_s = X \cup \{s\}$ is trivially closed. Indeed, in such case s will not participate in the dynamical operation of the reaction network $(X_s, \mathcal{R}_{X_s})$. Therefore, the dynamical analysis of a reaction network should always be reduced to the study of its reactive part [2].

Definition 10. *Let $\mathcal{C}\uparrow$ the set of reactive closed sets. We call $\mathcal{C}\uparrow$ the reactive closure structure.*

From now on we will refer to reactive closed sets simply as closed sets. Note that Lemma 2 does not apply to $\mathcal{C}\uparrow$.

Example 1. Consider the reaction network $(\{a, b, c\}, \{a \rightarrow b, a + c \rightarrow b + c\})$. For this reaction network we have that $\mathcal{C}\uparrow = \{\{a, b\}, \{a, b, c\}\}$, but $\{a, b\} \wedge_{CL} \{a, b, c\} = \{b\} \notin \mathcal{C}\uparrow$.

Algorithmic Remark 1. *The algorithm developed in [18] to compute a lattice structure of closed sets can turn extremely inefficient due to the addition of closed sets with non-reactive species. In particular, suppose $|\mathcal{M}| = n$ and $X \in \mathcal{C}\uparrow$ such that $|X| = m$, and suppose that the $n - m$ species not in X are not reactive with respect to X. In this case, the algorithm that computes the lattice structure of $\mathcal{C}$ will compute at least 2^{n-m} closed sets having non-reactive species.*

We now note some useful statements about the combination of reactive closed sets using $\vee_{CL}$ and $\wedge_{CL}$.

Lemma 3. *Let $X, Y \in \mathcal{C}\uparrow$, we have that*

1. $X \vee_{CL} Y \in \mathcal{C}\uparrow$
2. $X \wedge_{CL} Y \in \mathcal{C}\uparrow$ *if and only if $X \cap Y$ is reactive.*

Proof. Both assertions can be deduced directly from the fact that the generated closure of a reactive set is reactive.   □

Corollary 1. *$(\mathcal{C}\uparrow, \vee_{CL})$ is a join-semilattice.*

Corollary 1 illustrates that a constructive approach to compute $\mathcal{C}\uparrow$ can be safely done using $\vee_{CL}$. However, we will show that within $\mathcal{C}\uparrow$ there are also irrelevant closed sets.

3.2   Some Separable Closed Sets (such as the Incoherent and Separable) Are Irrelevant

The notion of separable closed set captures that a closed set can be decomposed into smaller closed sets in such a way that the dynamics of the set can be safely encoded by the dynamics of the parts.

Definition 11. *A closed set X is separable if and only if there exist $X_1 \neq X_2$ and $X_1, X_2 \subset X$ in $C\uparrow$ such that*

- $X_1 \cup X_2 = X$, *and*
- $\mathcal{R}_{X_1 \cup X_2} = \mathcal{R}_{X_1} \cup \mathcal{R}_{X_2}$.

If, additionally, we have that

- $\mathcal{R}_{X_1} \cap \mathcal{R}_{X_2} = \emptyset$ *we say X is incoherent.*
- $X_1 \cap X_2 = \emptyset$ *we say X is disconnected.*

For a separable set $X = X_1 \cup X_2$, its dynamical operation can be partitioned into the dynamical operation of X_1 and X_2. Thus, we can consider the combination $X_1 \vee_{CL} X_2 = X_1 \cup X_2 = X$ a trivial closet set. The notion of incoherent and disconnected reflect two stronger forms of separability because the decomposition of X into X_1 and X_2 is disjoint w.r.t the set of reactions and species respectively. However, these two possibilities do not cover all the ways in which a set is separable. Indeed, not all the separable sets are incoherent or disconnected.

Example 2. Let $\mathcal{M} = \{s_1, s_2, s_3, s_4, s_5, s_6\}$ and

$$\mathcal{R} = \{r_1 = s_1 \rightarrow 2s_1, r_2 = s_1 + s_2 \rightarrow s_3, r_3 = s_4 + s_5 \rightarrow s_3, r_4 = s_1 + s_6 \rightarrow 2s_6\},$$

and the following closed sets

$$X_1 = \{s_1\}, X_2 = \{s_1, s_2, s_3\}, X_3 = \{s_3, s_4, s_5\}, X_4 = \{s_1, s_6\}.$$

Note that

- $X_1 \cup X_3$ *is disconnected and thus incoherent and separable.*
- $X_2 \cup X_3$ *is separable and incoherent, but not disconnected because $X_2 \cap X_3 = \{s_3\}$,*
- $X_1 \cup X_4$ *is separable but neither incoherent or disconnected because $\mathcal{R}_{X_1} \cap \mathcal{R}_{X_4} = \{r_1\}$, and hence $X_1 \cap X_4 = \{s_1\}$.*

Definition 12. *Let $C\uparrow^{sep}$, $C\uparrow^{inc}$, and $C\uparrow^{dis}$, be the set of separable, incoherent and disconnected closed sets respectively.*

Lemma 4.
$$C\uparrow^{dis} \subseteq C\uparrow^{inc} \subseteq C\uparrow^{sep} \subseteq C\uparrow \subseteq C. \tag{4}$$

Definition 13. *Let $C\uparrow^{\downarrow} = C\uparrow - C\uparrow^{sep}$ the non-separable closure structure. We call a set $X \in C\uparrow^{\downarrow}$ non-separable.*

Non-separable reaction networks are the least trivial networks from a dynamical point of view. Indeed, their dynamics cannot be decomposed into the dynamics of any combination of its constituent parts. However, we will see that some separable reaction networks (that are necessarily not incoherent) are also non-trivial from a dynamical point of view. Indeed, the fact that X is separable into X_1 and X_2 (which implies that its dynamics can be explained in terms of the dynamics of X_1 and X_2) does not necessarily imply that, for other two sets X_1' and X_2' that also decompose X, we can decomposed the dynamics of X into the dynamics of X_1' and X_2'. So, we have that some decompositions of X do exhibit separability while other decompositions do not. Since the decomposition of X is done considering closed sets only, the number of decompositions which exhibit separability of X versus the total number of decompositions of X footprints its structural separability. In particular, we will show that there are different degrees of separability, ranging from non-separability to full separability, and that such degree can be characterized by a suitable notion of synergy for reaction networks.

4   On Separability and Synergy

4.1   The Notion of Synergy

A synergy in our setting corresponds to the appearance of novel reactions when closed sets are combined.

Definition 14. *Let $\boldsymbol{X} = \{X_1, ..., X_k\}$ be a collection of closed sets. $\boldsymbol{X}$ is a synergy if and only if*

1. $\mathcal{R}_{\boldsymbol{X}} \supset \bigcup_{X \in \boldsymbol{X}} \mathcal{R}_X$, and

2. for any sub-collection $\boldsymbol{X}'$ of $\boldsymbol{X}$ we have that $\mathcal{R}_{\boldsymbol{X}'} = \bigcup_{X \in \boldsymbol{X}'} \mathcal{R}_X$.

Given a set X, the existence of a collection $\boldsymbol{X}$ of subsets of X that is a synergy represents a necessary condition for the non-separability of X. Note that condition (i) indicates that the combination of sets in the collection triggers a 'novel' reaction, i.e. that cannot be triggered by any of the former sets in the collection, and (ii) indicates that such novel reaction is generated in a minimal way, i.e. no sub-collection of sets is able to trigger the novel reaction. The notion of separability is intimately related to the notion of synergy in a reaction network. The following lemma, although trivial, will be helpful to prove some important statements about collections of sets that have a synergy.

Lemma 5. *Let $\boldsymbol{X} = \{X_1, ..., X_k\}$ be a collection of sets that is synergy. Then, there exist a reaction $r \in \mathcal{R}_{\boldsymbol{X}}$ such that*

- *$r \notin \mathcal{R}_{\boldsymbol{X}'}$ for every sub-collection $\boldsymbol{X}'$ of $\boldsymbol{X}$.*
- *$supp(r) \cap X_i \neq \emptyset$, for $i = 1, ..., k$.*

Proof. To prove the first statement: By condition 1 of Definition 14 we have that $\mathcal{R}_\mathbf{X} \supset \bigcup_{X \in \mathbf{X}} \mathcal{R}_X$. This means that there is a reaction $r \in \mathcal{R}_\mathbf{X}$ which is not in $\bigcup_{X \in \mathbf{X}} \mathcal{R}_X$.

To prove the second statement (by contradiction), suppose that there exist $i = j$ such that $\mathrm{supp}(r) \cap X_j = \emptyset$, and let $\mathbf{X}' = \mathbf{X} - \{X_j\}$. Then, we have that $r \in \mathcal{R}_{\mathbf{X}'}$, and this implies that $\mathcal{R}_{\mathbf{X}'} \supset \bigcup_{X \in \mathbf{X}'} \mathcal{R}_X$, which contradicts condition 2 of Definition 14. □

Corollary 2. *Let $\mathbf{X} = \{X_1, ..., X_k\}$ be a collection of sets that is a synergy. Then, there exists a closed set $Z \subseteq G_{CL}(\mathbf{X})$ such that for $i = 1, ..., k$*

- $Z \cap X_i \neq \emptyset$, *and*
- $Z \not\subseteq X_i$.

Proof. By Lemma 5 we have that there is $r \in \mathcal{R}_\mathbf{X}$ such that $r \bigcup_{X \in \mathbf{X}} \mathcal{R}_X$. Then, both statements trivially follow by defining $Z = G_{CL}(\mathrm{supp}(r))$. □

Lemma 5 and Corollary 2 show that given a collection $\mathbf{X}$ that is a synergy, their closure triggers a new reaction r which in turn produces a new closed set that is not contained in any of the former sets in $\mathbf{X}$.

Corollary 3. *Let X be a closed set. If the number of closed sets that are contained in X is less than three, then there is no collection of subsets of X that is a synergy.*

Proof. We will show that if X has one or two closed sets contained in it, then it is not possible to construct a collection of subsets of X that is a synergy. First, suppose that there is only one closed set $X_1 \subset X$. Then, $\mathbf{X} = \{X_1\}$ clearly is not a synergy. Now, suppose that there are two sets $X_1, X_2 \subset X$ and no other subset of X is closed. Then, clearly neither $\mathbf{X} = \{X_1\}$ nor $\mathbf{X} = \{X_2\}$ is a synergy. Finally, if $\mathbf{X} = \{X_1, X_2\}$ is a synergy, then by Lemma 2 there is a closed set $Z \subset X$ such that $Z \neq X_1$ and $Z \neq X_2$ which entails a contradiction to the fact that X contains only two closed sets. □

The following definition characterizes the reactions that reflect synergies within a reaction network.

Definition 15. *Let $\mathbf{X} = \{X_1, ..., X_k\}$ be a collection of sets that is a synergy. Then, we define $\mathcal{R}_\mathbf{X}^{syn}$ the set of reactions $r \in \mathcal{R}_\mathbf{X}$ such that $r \notin \bigcup_{X \in \mathbf{X}} \mathcal{R}_X$. We call $\mathcal{R}_\mathbf{X}^{syn}$ the set of synergetic reactions of $\mathbf{X}$, and for each $r \in \mathcal{R}_\mathbf{X}^{syn}$ we call $G_{CL}(\mathrm{supp}(r))$ a synergetic closure of $\mathbf{X}$.*

4.2   Computational Tractability of Synergies - an Interlude

Although the results about synergetic reactions and synergetic closures (Lemma 5 and Corollary 2) are easy to prove, identifying that a collection of sets $\mathbf{X}$ is a synergy is not trivial. In particular, the verification of condition 2. of Definition 14 requires to check that every novel reaction $r \in \mathcal{R}_{\mathbf{X}}$ cannot be triggered by any sub-collection $\mathbf{X}'$ of $\mathbf{X}$. This implies that, in principle, one must check that r cannot be triggered by any of the $2^{|\mathbf{X}|} - 1$ sub-collections that can be generated from $\mathbf{X}$. The latter suggests that identifying synergies could lead to a combinatorial explosion. However, the following lemma shows that the problem is tractable.

Lemma 6. *Let* $\rho = \max\{|supp(r)|, r \in \mathcal{R}\}$. *Then, a collection of sets* $\mathbf{X}$ *that is a synergy implies* $|\mathbf{X}| \leq \rho$.

Proof. A simple sketch of the proof is as follows: If $\mathbf{X}$ is a synergy then each set $X_i \in \mathbf{X}$ must contribute with at least one species to trigger the synergetic reaction r. Then, it is evident that the collection $\mathbf{X}$ cannot have more elements than the maximum of the number of reactants of the reactions in $\mathcal{R}$, which is bounded by ρ.

Formal proof: Assume there is a collection of sets $\mathbf{X}$ that is a synergy and $|\mathbf{X}| > \rho$. Without loss of generality let $|\mathbf{X}| = \rho + 1$. Let $\mathbf{X} = \{X_1, ..., X_{\rho+1}\}$, and let $r \in \mathcal{R}_{\mathbf{X}}^{syn}$. Since $|supp(r)| \leq \rho$, we can denote $supp(r) = \{s_{r_1}, ...s_{r_k}\}$ with $k \leq \rho$. Now let $X_{r_i} = \{X_i \in \mathbf{X}/s_{r_i} \in X_i\}$ for $i = 1, ..., k$. $\mathbf{X}_{r_i}$ denotes all the sets in $\mathbf{X}$ that contain s_{r_i}. Now, let X_i' be the first set of the collection $\mathbf{X}_{r_i}$, for $i = 1, ..., k$, and $\mathbf{X}' = \{X_1', ..., X_k'\}$. By construction we have that $\mathcal{R}_{\mathbf{X}'} \supset \bigcup_{i=1}^{k} \mathcal{R}_{X_i}$, which contradicts condition 2 of Definition 14. $\square$

Lemma 6 is an important result to understand the inner structure of closed reaction networks. On the one hand, it shows that the number of closed sets in a collection required to be a synergy is bounded by ρ, which provides a simple necessity criteria for a synergy and shows that the verification of synergies is tractable. On the other hand, it suggests that if one is able to identify a collection of sets $\mathcal{B}$ from which all synergies are obtained from sub-collections of $\mathcal{B}$, then the number of collections of sets that is a synergy, and hence the number of synergetic closures, is bounded by

$$\sum_{i=2}^{\rho} |\mathcal{B}|^i = \frac{|\mathcal{B}|^{\rho+1} - |\mathcal{B}|^2}{|\mathcal{B}| - 1}. \tag{5}$$

We will elucidate which closed sets can be used to build $\mathcal{B}$. To do so, we need to clarify the formal relation between synergies and non-separable sets.

4.3   Non-separability Means Full Synergy

Lemma 7. *Let* X *be a closed set such that for every* $X_1 \neq X_2$ *closed sets in* X *either* $X_1 \subset X_2$ *or* $X_2 \subset X_1$. *Then, there exist* r *in* R_X *such that* $X = G_{CL}(supp(r))$.

Proof. Since all pairs of closed sets in X are related by inclusion, let $\mathbf{X} = \{X_1, ..., X_k = X\}$ be the set of all closed sets contained in X with $X_i \subset X_{i+1}$, $i = 1, ..., k - 1$. Now, suppose there is no r such that $G_{CL}(\text{supp}(r)) = X$. Then, there are two reactions $r', r'' \in \mathcal{R}_X$ such that $G_{CL}(\text{supp}(r')) = X' \not\subset G_{CL}(\text{supp}(r_2)) = X''$ and $G_{CL}(\text{supp}(r'')) = X'' \not\subset G_{CL}(\text{supp}(r_1)) = X'$. Since X' and X'' are closed sets, the latter entails a contradiction. $\square$

Lemma 8. *Let X be a non-separable closed set. Let $r \in \mathcal{R}_X$, and $G_{CL}(supp(r)) = Z$. Then,*

1. *$Z = X$, or*
2. *$G_{CL}(X - Z) = X$.*

Proof. If $Z = X$ then condition *1.* holds. If $Z \neq X$, then we will prove that $Y = G_{CL}(X - Z)$ is equal to X and thus condition *2.* holds. First note that $Y \supset X - Z$, and thus $Y \cup Z = X$. Now, suppose that $Y \neq X$. The latter implies that $r \notin \mathcal{R}_Y$ because otherwise Y would be able to trigger r and as consequence $Z \subseteq Y$ which would imply $Y = X$. Hence, let $\bar{Y} = X - \text{supp}(r)$. Clearly $r \notin \mathcal{R}_{\bar{Y}}$. Moreover, note that $\bar{Y}$ is closed because otherwise it will have to trigger one new reaction but the only reactions in X and not in $\bar{Y}$ are those that can be triggered with $\text{supp}(r)$ which is not in $\bar{Y}$. As a consequence we have that $\mathcal{R}_{\bar{Y}} \cup \mathcal{R}_Z = \mathcal{R}_X$.

Therefore, we have that $\bar{Y} \neq Z$, $\bar{Y} \cup Z = X$ and $\mathcal{R}_{\bar{Y}} \cup \mathcal{R}_Z = \mathcal{R}_X$, which implies X is separable (and this entails a contradiction to the fact that $Y \neq X$). $\square$

Theorem 2. *Ley X be a non-separable set, then one of two alternatives hold:*

1. *There exist $r \in \mathcal{R}_X$ such that $G_{CL}(supp(r)) = X$,*
2. *There does not exist $r \in \mathcal{R}_X$ such that $G_{CL}(supp(r)) = X$, and for every $\mathbf{X} = \{X_1, X_2\}$ such that $X_1, X_2 \subset X$ and $X_1 \vee_{CL} X_2 = X$, $\mathbf{X}$ is a synergy.*

Proof. Since X is non-separable, condition *1* follows from condition *1* of Lemma 8. We now prove that condition *2* of Lemma 8 implies condition *2* of our theorem:
Since $X_1, X_2 \subset X$ and $X_1 \vee_{CL} X_2 = X$, we have by Lemma 8 that $G_{CL}(X - X_1) = G_{CL}(X - X_2) = X$ because otherwise X would be separable. Hence we have that

$$\mathcal{R}_{X_2} \subset \mathcal{R}_{G_{CL}(X-X_2)} = \mathcal{R}_X, \text{ and } \mathcal{R}_{X_1} \subset \mathcal{R}_{G_{CL}(X-X_1)} = \mathcal{R}_X.$$

Now, let $\mathcal{R}_1 = \mathcal{R}_X - \mathcal{R}_{X_1}$ and $\mathcal{R}_2 = \mathcal{R}_X - \mathcal{R}_{X_2}$. Note that $\mathcal{R}_1 \cap \mathcal{R}_2$ cannot be the empty set because that would imply that every reaction in $\mathcal{R}_1 \cap \mathcal{R}_2$ must be either in X_1 or X_2, which would imply X is separable. Therefore, there is a reaction $r \in \mathcal{R}_X$ in that is not in $\mathcal{R}_{X_1} \cup \mathcal{R}_{X_2}$. Hence, $\mathbf{X} = \{X_1, X_2\}$ is a synergy. $\square$

In Fig. 1 we show an example of a non-separable set, and a separable set that has a synergy.

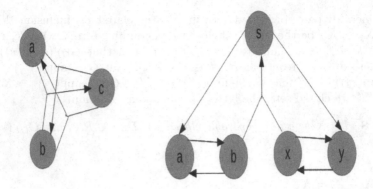

Fig. 1. Left: example of a non-separable network. Right: example of separable network with synergy

Theorem 2 establishes the formal relation between non-separable sets and collections of sets that have a synergy. In particular, it shows that there are two basic types of non-separable sets. The first type, which we will call *basic sets* are the closure of a single reaction. We know that a reaction network $(\mathcal{M}, \mathcal{R})$ has at most $|\mathcal{R}|$ basic sets. The second type, are sets X such that any pair of sets $\mathbf{X} = \{X_1, X_2\}$ whose union generates X via closure, necessarily is a synergy.

Definition 16. *Let* $X \in \mathcal{C}\!\uparrow$, *and let* $\mathcal{D}(X) = \{\mathbf{X}^1 = \{X_1^1, X_2^1,\}..., \mathbf{X}^k = \{X_1^k, X_2^k\}\}$ *be the collection of pairs of closed subsets of* X *such that* $X_1^i \vee_{CL} X_2^i = X$, *for* $i = 1, ..., k$. *We call* $\mathcal{D}(X)$ *the separability decomposition of* X. *In addition, let* j *be the number of elements in* $\mathcal{D}(X)$ *that are a synergy. We say* X *is* ρ-*separable, with* $\rho = 1 - \frac{j}{k}$.

Lemma 9. *X is non-separable if and only if X is* $0-separable$

5   Discussion and Conclusion

We have introduced the notions of separability and synergy for reaction networks and how they are related. In particular, the dynamics of a reaction network with no synergies, i.e. $1-$separable, can be safely studied from the dynamics of its parts, while for the opposite case, the so-called non-separable reaction networks ($0-$separable), there is no decomposition that can be used to study its dynamics. The intermediate case, i.e. sets that are $\rho-$separable with $0 < \rho < 1$ can be decomposed only in certain ways that ensure the possibility to study their dynamical operation. It is remarkable that the so-called non-separable closed sets entail structurally 'entangled' entities in the reaction network. Indeed, if one removes any part of it, the closure of the remaining part will become either the non-separable set again, or the empty set.

Although we have not provided a specific algorithm to compute closed sets, the results of this article provide a theoretical ground to develop an algorithm which (i) is able to efficiently build from bottom-up the set of closed sets, (ii) is able to retrieve decompositions of closed sets of the upper levels that can be applied to simplify dynamical analysis, and (iii) is able to identify non-separable sets. Indeed, from Theorem 2 it is derived that basic sets (the first type of sets in Theorem 2) entail a sufficient collection of minimal closed sets such that their combinations can be used to compute all dynamically relevant closed sets. Hence, such bottom-up algorithm must first compute the basic sets and then combine them appropriately. Fortunately though, Lemma 6 helps us to bound the number of operations needed to compute all the closed sets that have synergetic closures. Indeed, considering B to be the set of basic sets, we have a bound for number of synergetic closures in Eq. 5.

Another important aspect of our approach becomes clear when one reverts the question we have been asking. So, instead of thinking how do we decompose a non-trivial reaction network (which is what we did in this article), we could ask what are the combinations of reaction networks that lead to non-trivial reaction networks? The answer to this question is the key for the development of an efficient algorithm to compute the relevant closed sets, and it is currently a matter of investigation. An element that might be of help is to observe how such combinations appear in real reaction networks. Namely, there might be some regularities in the way that large dynamically relevant networks are formed from synergies of smaller parts. We believe that this exploration could give some important clues to understand the structure of (dynamically relevant) large reaction networks.

In addition, we plan to extend our results combining Theorem 2 with the decomposition theorem for organizations [20,23] so an efficient algorithm that computes the set of organizations of a reaction network, i.e. avoiding all combinatorial problems identified in previous algorithms [2,18], can be developed. It is the opinion of the authors that the structural study of reaction networks will provide important breakthroughs from an algorithmic and applied point of view. Specially considering that reaction networks have the capacity to model not only biochemical but also systems of a wide range of types [22].

References

1. Centler, F., di Fenizio, P.S., Matsumaru, N., Dittrich, P.: Chemical organizations in the central sugar metabolism of *Escherichia Coli*. In: Deutsch, A., Brusch, L., Byrne, H., Vries, G., Herzel, H. (eds.) Mathematical Modeling of Biological Systems, Volume I. MSSET, pp. 105–119. Birkhäuser, Boston (2007). https://doi.org/10.1007/978-0-8176-4558-8_10
2. Centler, F., Kaleta, C., di Fenizio, P.S., Dittrich, P.: Computing chemical organizations in biological networks. Bioinformatics **24**(14), 1611–1618 (2008)
3. Centler, F., Kaleta, C., Speroni di Fenizio, P., Dittrich, P.: A parallel algorithm to compute chemical organizations in biological networks. Bioinformatics **26**(14), 1788–1789 (2010)
4. Dittrich, P., Winter, L.: Chemical organizations in a toy model of the political system. Adv. Complex Syst. **11**(04), 609–627 (2008)
5. Dittrich, P., Di Fenizio, P.S.: Chemical organisation theory. Bull. Math. Biol. **69**(4), 1199–1231 (2007)

6. Garg, V.K.: Introduction to Lattice Theory with Computer Science Applications. Wiley, Hoboken (2015)
7. Horn, F., Jackson, R.: General mass action kinetics. Arch. Ration. Mech. Anal. **47**(2), 81–116 (1972)
8. Kreyssig, P., Escuela, G., Reynaert, B., Veloz, T., Ibrahim, B., Dittrich, P.: Cycles and the qualitative evolution of chemical systems. PLoS ONE **7**(10), e45772 (2012)
9. Kreyssig, P., Wozar, C., Peter, S., Veloz, T., Ibrahim, B., Dittrich, P.: Effects of small particle numbers on long-term behaviour in discrete biochemical systems. Bioinformatics **30**(17), i475–i481 (2014)
10. Matsumaru, N., Centler, F., di Fenizio, P.S., Dittrich, P.: Chemical organization theory applied to virus dynamics (Theorie chemischer organisationen angewendet auf infektionsmodelle). Inf. Technol. **48**(3), 154–160 (2006)
11. Matsumaru, N., Centler, F., di Fenizio, P.S., Dittrich, P.: Chemical organization theory as a theoretical base for chemical computing. In: Proceedings of the 2005 Workshop on Unconventional Computing: From Cellular Automata to Wetware, pp. 75–88. Luniver Press (2005)
12. Matsumaru, N., Lenser, T., Hinze, T., Dittrich, P.: Toward organization-oriented chemical programming: a case study with the maximal independent set problem. In: Dressler, F., Carreras, I. (eds.) Advances in Biologically Inspired Information Systems. SCI, vol. 69, pp. 147–163. Springer, Heidelberg (2007). https://doi.org/10.1007/978-3-540-72693-7_8
13. Peter, S., Veloz, T., Dittrich, P.: Feasibility of organizations – a refinement of chemical organization theory with application to P systems. In: Gheorghe, M., Hinze, T., Păun, G., Rozenberg, G., Salomaa, A. (eds.) CMC 2010. LNCS, vol. 6501, pp. 325–337. Springer, Heidelberg (2010). https://doi.org/10.1007/978-3-642-18123-8_25
14. Peter, S., Dittrich, P.: On the relation between organizations and limit sets in chemical reaction systems. Adv. Complex Syst. **14**(01), 77–96 (2011)
15. Reddy, V.N., Mavrovouniotis, M.L., Liebman, M.N.: Petri net representations in metabolic pathways. In: ISMB, vol. 93, pp. 328–336, July 1993
16. Schuster, S., Hilgetag, C.: On elementary flux modes in biochemical reaction systems at steady state. J. Biol. Syst. **2**(02), 165–182 (1994)
17. Schilling, C.H., Schuster, S., Palsson, B.O., Heinrich, R.: Metabolic pathway analysis: basic concepts and scientific applications in the postgenomic era. Biotechnol. Prog. **15**(3), 296–303 (1999)
18. Speroni di Fenizio, P.: The lattice of chemical organisations. In: Artificial Life Conference Proceedings 13, pp. 242–248. MIT Press, Cambridge, July 2015
19. Strogatz, S.H.: Nonlinear Dynamics and Chaos: With Applications to Physics, Biology, Chemistry, and Engineering. CRC Press, Boca Raton (2018)
20. Veloz, T.: A computational study of Algebraic Chemistry (Master's thesis) (2010)
21. Veloz, T., Razeto-Barry, P., Dittrich, P., Fajardo, A.: Reaction networks and evolutionary game theory. J. Math. Biol. **68**(1–2), 181–206 (2014)
22. Veloz, T., Razeto-Barry, P.: Reaction networks as a language for systemic modeling: fundamentals and examples. Systems **5**(1), 11 (2017a)
23. Veloz, T., Razeto-Barry, P.: Reaction networks as a language for systemic modeling: on the study of structural changes. Systems **5**(2), 30 (2017b)
24. Veloz, T., Ramos, R., Maldonado, P., Moisset, P.: Reaction networks as a tool for representing and analyzing ecological networks (2018, in preparation)

A Logical Framework for Modelling Breast Cancer Progression

Joëlle Despeyroux[1], Amy Felty[2], Pietro Liò[3], and Carlos Olarte[4(✉)]

[1] INRIA and CNRS, I3S Laboratory, Sophia-Antipolis, France
[2] School of Electrical Engineering and Computer Science, University of Ottawa, Ottawa, Canada
[3] Department of Computer Science and Technology, University of Cambridge, Cambridge, UK
[4] ECT, Universidade Federal do Rio Grande do Norte, Natal, Brazil
carlos.olarte@gmail.com

Abstract. Data streams for a personalised breast cancer programme could include collections of image data, tumour genome sequencing, likely at the single cell level, and liquid biopsies (DNA and Circulating Tumour Cells (CTCs)). Although they are rich in information, the full power of these datasets will not be realised until we develop methods to model the cancer systems and conduct analyses that transect these streams. In addition to machine learning approaches, we believe that logical reasoning has the potential to provide help in clinical decision support systems for doctors. We develop a logical approach to modelling cancer progression, focusing on mutation analysis and CTCs, which include the appearance of driver mutations, the transformation of normal cells to cancer cells in the breast, their circulation in the blood, and their path to the bone. Our long term goal is to improve the prediction of survival of metastatic breast cancer patients. We model the behaviour of the CTCs as a transition system, and we use Linear Logic (LL) to reason about our model. We consider several important properties about CTCs and prove them in LL. In addition, we formalise our results in the Coq Proof Assistant, thus providing formal proofs of our model. We believe that our results provide a promising proof-of-principle and can be generalised to other cancer types and groups of driver mutations.

1 Introduction

Cancer is a complex evolutionary phenomenon, characterised by multiple levels of heterogeneity (inter-patient, intra-patient and intra-tumour), multiscale events (i.e. changes at intracellular, intercellular, tissue levels), multiomics variability (i.e. changes to chromatine, epigenetic and transcriptomic levels) that affect all aspects of clinical decisions and practice. The most remarkable phenomenon is the occurrence of numerous somatic mutations, of which only a subset contributes to cancer progression. The dynamic genetic diversity, coupled with epigenetic plasticity, within each individual cancer induces new genetic architectures and clonal evolutionary trajectories.

© Springer Nature Switzerland AG 2019
M. Chaves and M. A. Martins (Eds.): MLCSB 2018, LNCS 11415, pp. 121–141, 2019.
https://doi.org/10.1007/978-3-030-19432-1_8

A striking feature of cancer is the subclonal genetic diversity, i.e., the presence of clonal succession and spatial segregation of subclones in primary sites and metastases. At the root of the subclonal expansion there are developmentally regulated potentially self-renewing cells. After initiation, multiple subclones often coexist, signalling parallel evolution with no selective sweep or clear fitness advantage that becomes evident with therapies. Fitness calculation for each subclonal is difficult as subclones can form ecosystems and can cooperate through paracrine loops or interact through stromal, endothelial and inflammatory cells.

Tumour stages describe the progress of the tumour cells. One widely adopted approach is the American Joint Committee on Cancer (AJCC) tumour node metastasis (TNM) staging system. It classifies tumours with a combined stage between I-IV using three values: T gives the size of the primary tumour and extent of invasion, N describes if the tumour has spread to regional lymph nodes and M is indicative of distant metastasis. In terms of prognosis, Stage I patients have the best prognosis with 5 year survival rates (80–95%). The survival rates progressively worsen with each stage. Even with advances in targeted therapies, Stage IV patients have survival rates of just over two years. The integration of blood tests, biopsies, medical imaging, with genomic data have allowed the classification of many subtypes of cancers with striking differences of driver mutations and survival patterns. Therefore, the current data stream goals for a personalised breast cancer program should include the generation of tumour whole genome sequencing (DNA and RNA) for patients with breast cancer. Another data stream goal could focus on liquid biopsies. This will consist of data obtained from circulating tumour DNA (ctDNA) and single cell analysis. However, the full power of these datasets will not be realised until we leverage advanced statistical, mathematical and computational approaches to devise the needed procedures to conduct analyses that transect these streams.

There is a very rich body of biomedical statistics, machine learning and epidemiological literature for cancer data analysis which includes methods ranging from survival analysis, i.e., the effect of a risk factor or treatment with respect to cancer progression, analyses of co-alteration and mutual exclusivity patterns for genetic alterations, gene expression analyses, to network science algorithms (see e.g., [4,5,8,17,19,26,33]). For example, in survival modelling, the data is referred to as the time to event date and the objective is to analyse the time that passes before an event occurs due to one or more covariates [22,23]. We believe that together with machine learning and biostatistics, there is a role for a logical approach in guiding optimal treatment decisions and in developing a risk stratification and monitoring tool to manage cancer. In this work, we focus on the use of a formal logical framework (as described below) to provide a reliable hypothesis-driven decision making system based on molecular data.

1.1 Formal Methods for Systems Biology

Computational systems biology provides a variety of methods for understanding the structure of biological systems and for studying their dynamics. In order to capture the *qualitative* nature of dynamics, approaches include Petri nets [9],

π-calculus [31], bio-ambients [30], and rule-based modelling languages such as Biocham [18] and Kappa [11]. Molecular Logic [37] uses boolean logic gates to define regulations in networks. One of the most successful approaches to model and analyse signal transduction networks, inside the cells, is Pathway Logic [35] (a system based on rewriting rules). Our approach, based on logical frameworks, can be used to model biological networks both inside and outside the cell.

The dynamics from a *quantitative* point of view can be captured by means of ordinary or stochastic differential equations. More recent approaches include hybrid Petri nets [21] and hybrid automata [1], piecewise linear equations [24], stochastic π-calculus [29], and rule-based languages with continuous/stochastic dynamics such as Kappa [11], Biocham [18], or BioNetGen [7].

One of the most common approaches to the formal verification of biological systems is model checking [10] that exhaustively enumerates all the states reachable by the system. In order to apply such a technique, the biological system should be encoded as a finite transition system and relevant system properties should be specified using temporal logic.

1.2   Logical Frameworks

In contrast to the aforementioned approaches, we encode both biological systems and temporal properties in logic, and prove that the properties can be derived from the system. This approach is new, only proposed in two previous works up to now [12,28]. In the present work, we choose discrete modelling, with temporal transition constraints. We believe that discrete modelling is crucial in systems biology since it allows taking into account some phenomena that have a very low chance of happening (and could thus be neglected by differential approaches), but which may have a strong impact on system behaviour.

We advocate the use of logical frameworks as an *unified and safe* approach to both specifying and analysing biological systems. *Logical Frameworks* are logics designed to formally study a variety of systems. The formal study of such systems means providing both formal models and proofs of properties of the systems. In the case where the logical systems are themselves logics, the logical framework enables the proofs of both *meta-theoretical* theorems (about the logic being formalised) and *object level* theorems (about the systems being encoded in the formalised logic). We shall use *Linear Logic* (LL) [20] as the intermediate logic, formalised in the Calculus of Constructions, which is a type theory implemented in the Coq Proof Assistant [6]. We note that the Coq system has been (partially) proven correct in itself, extensive meta-theoretical studies of LL are available in the Coq system (see e.g., [38]), and our encoding of biological systems is *adequate* (Sect. 3.1). This means that we prove that the formal model of the system correctly encodes the intended biological system. The approach is thus an unified approach, fully based on logic, and a safe approach, as each step is proved correct, as far as it can be.

We leverage our formalisation of LL in Coq to give a natural and direct characterisation of the state transformations of Circulating Tumour Cells (CTCs). For instance, a rule describing the evolution of a cell n, in a region r, from a healthy cell (with no mutations) to a cell that has acquired a mutation TGFβ can be modelled by the linear implication $C(n, r, [])$ $\multimap$ $C(n, r, [TGF\beta])$. This formula describes the fact that a state where a cell $C(n, r, [])$ is present can evolve into a state where $C(n, r, [TGF\beta])$ holds. More interestingly, the LL specification can be used to prove some desired properties of the system. For instance, it is possible to prove *reachability properties*, i.e., whether the system can reach a given state (Sect. 4.1) or even more abstract (meta-level) properties such as checking *all* the possible evolution paths the system can take under certain conditions (Sect. 4.2). Finally, we attain a certain degree of automatisation in our proofs which opens the possibility of testing recent proposed hypotheses in the literature.

Organisation. The rest of the paper is organised as follows. Section 2 describes the most relevant properties related to cancer mutations and CTCs which we believe are key factors driving the model dynamics. In Sect. 3 we recall the theory of LL and we specify in it the dynamics of CTCs. We prove correct our model in Theorem 1 by showing that transitions in the system are in one-to-one correspondence with logical steps. Reachability and meta-level properties of the system are proved in Sect. 4. Finally we conclude with a discussion on challenges and opportunities of logical frameworks in cancer studies. There is a companion appendix with the proofs of the results presented here. Moreover, all the proofs of the properties of our model were certified in Coq and are available at http://subsell.logic.at/bio-CTC/.

2   Tumour Cells in Metastatic Breast Cancer

In this section we first describe the mutations involved in cancer in general and then, we focus on the evolution of circulating tumour cells described in this work.

Cancer Mutations. Cancer mutations can be divided into drivers and non-drivers (or passengers). The accumulation of evidence of clonal heterogeneity and the observation of the arrival of drug resistance in clonal sub-populations suggest that mutations usually classified as non-drivers may have an important role in the fitness of the cancer cell and in the evolution and physiopathology of cancer. Similarly, mutations that alter the metabolism and the epigenetics may modify the fitness of the cancer cells. A meaningful way to identify drivers and passenger mutations is to use a statistical estimator of the impact of mutations such as FATHMM-MKL and a very large mutation database such as Cosmic (http://cancer.sanger.ac.uk/) [34].

The mutation process (causing tumour evolution) generates intra-tumour heterogeneity. The subsequent selection and Darwinian evolution (including immune escape) on intra-tumour heterogeneity is the key challenge in cancer medicine. Those clones that have progressed more than the others will have larger influence on patient survival. The amount of heterogeneity can vary between zero and up to over several thousand mutations, found to be heterogeneous within primary tumours or between primary and metastatic. The heterogeneity could be investigated through molecular biology techniques such as single cell sequencing, in situ PCR and could also be phenotypically classified using microscope image analysis. In case of large heterogeneity we could assume that the survival of the patient strongly depends on the mutations of the most aggressive clones/cells.

Circulating Tumour Cells. We follow the study of the evolution of Circulating Tumour Cells in metastatic breast cancer in [4], where the authors use differential equations. This reference has an extensive discussion on the modelling choices, in particular concerning the driver mutations.

In [4] the probability for a cell in a duct in the breast to metastasise in the bone depends on the following mutational events:

1. A mutation in the TGFβ pathways frees the cell from the surrounding cells.
2. A mutation in the EPCAM gene makes the cell rounded and free to divide. Then the cell enters the blood stream and becomes a circulating cancer cell.
3. In order to survive, this cell needs to over express the gene CD47 that prevents attacks from the immune system.
4. Finally, there are two mutations that allow the circulating cancer cell to attach to the bone tissue and start the deadly cancer there: CD44 and MET.

Hence, a cancer cell has four possible futures: (a) acquiring a driver mutation; (b) acquiring a passenger mutation which does not cause too much of a viability problem: it simply increases a sort of "counter to apoptosis"; (c), the new (i.e., last) mutation brings the cell to apoptosis; and (d), moving to the next compartement, or seeding in the bone.

The behaviour of the cells depends on the *compartments* the cells live in (here the breast, the blood and the bone), the other cells (i.e., the *environment*: the availability of food/oxygen or the pressure by the other cells), and the behaviour of the surface proteins (the *mutations*). In this work, we shall formalise the compartments and the mutations, and leave the formalisation of the environment to future work. Note that this environment plays a role only in the breast.

The *phenotype* of a cell is characterised by both the number of its mutations and its fitness. In biology the *fitness* is the capability of the cell to survive and produce offsprings. The cell's viability is particularly dependent on metabolic health and energy level. Most of the cell metabolic health depends on the accumulation of mutations that affect the production of enzymes involved in catalysing energetic expensive reactions and cell homeostasis. The fitness is particularly altered by the occurrence of driver mutations: each driver mutation provides the cell with additional fitness. Non-driver mutations, on the other hand, may accumulate in large numbers, and may affect cell stress response due to the altered metabolism and the competition with other neighboring cells [15]. Wet-lab tests for cell fitness and stress responsiveness have been recently developed, see for instance [3,32,36]. In our formalisation, the fitness will be a parameter of the cells. Physicians see the appearance of the cell (round, free, etc.), while biologists see the mutations. Our model can take both into account.

We extend the model in [4] with a few rules modelling DNA repair - of passenger mutations. These rules, only available for cells with TGFβ or EPCAM mutations, (i.e. before CD47 mutation), represent DNA repair by increasing the fitness by one. Note that this addition introduces *cycles* in our model. (i.e., it is possible for a cell to go back to a previous state).

3   Specification in Linear Logic

Linear Logic (LL) [20] is a *resource conscious* logic particularly well suited for describing state transition systems. LL has been successfully used to model such diverse systems as planning, Petri nets, process calculi, security protocols, multiset rewriting, graph traversal algorithms, and games. In this section we formalise in LL the behaviours of the Circulating Tumour Cells and we prove this formalisation correct. In each step of the formalisation, we shall give an intuitive description of the LL connectives and their proof rules that should suffice to understand our developments. The reader may find in [20] a more detailed account on the proof theory of LL.

LL is a substructural logic where there is an explicit control over the number of times a formula can be used in a proof. Formulas can be split into two sets: classical (those that can be used as many times as needed) or linear (those that are consumed after being used). Using a dyadic system for LL [2], sequents take the form $\Gamma \; ; \; \Delta \vdash G$ where G is the formula (goal) to be proved (examples in Sect. 4), Γ is the set of classical formulas and Δ is the multiset of linear formulas. We store in Γ the formulas representing the rules of the system and in Δ the atomic predicates representing the state of the system, namely:

- $C(\mathbf{n}, \mathbf{c}, \mathbf{f}, \mathbf{lm})$, denoting a cell n (a natural number used as an id), in a compartment c (breast, blood, or bone), with a phenotype given by a *fitness* degree $f \in 0 \mathrel{..} 12$ and a list of driver mutations lm. The list of driver mutations lm is built up from mutations TGFβ, EPCAM, CD47, CD44, and/or MET, to which we add **seeded**, for the cells seeded in the bone. As TGFβ is required before any further mutations, a list of mutations [EPCAM, ...] will by convention mean [TGFβ, EPCAM, ...];
- $A(\mathbf{n})$, representing the fact that the cell n has gone to apoptosis;
- and $T(\mathbf{t})$, stating that the current time-unit is t.

The linear implication $F_1 \multimap F_2$ models a state transformation where F_1 is consumed to later produce F_2. The proof rules are:

$$\frac{\Gamma; \Delta, F_1 \vdash F_2}{\Gamma; \Delta \vdash F_1 \multimap F_2} \, {\multimap}R \qquad \frac{\Gamma; \Delta_1 \vdash\Downarrow F_1 \quad \Gamma; \Delta_2, \Downarrow F_2 \vdash G}{\Gamma; \Delta_1, \Delta_2, \Downarrow F_1 \multimap F_2 \vdash G} \, {\multimap}L$$

In ${\multimap}R$, the proof of $F_1 {\multimap} F_2$ requires the use of the resource F_1 to conclude F_2. This rule is invertible (i.e., the premise is provable if and only if the conclusion is provable). Hence, this rule belongs to the *negative* phase of the construction of a proof, where, without losing provability, we can apply all the invertible rules in any order. The rule ${\multimap}L$ shows the resource awareness of the logic: part of the context (Δ_1) is used to prove F_1 and the remaining resources (Δ_2) must be used to prove G. The classical context Γ is not divided but copied in the premises. The rule ${\multimap}L$ is non-invertible and then, it belongs to the *positive* phase. The notation $\Downarrow F_1 \multimap F_2$ means that we *decide* to *focus on* that formula and then, we have to keep working on the subformulas F_1 and F_2 (notation $\Downarrow F_1$ and $\Downarrow F_2$).

Each rule of the biological system is associated with a delay (see the terms of the form d_i in Fig. 1), which depends on the fitness parameter. Such parameters decrease marginally with passenger mutations and increase drastically with driver mutations. A typical rule in our model is then as follows:

$$\texttt{rl}(br_{e0.1}) \stackrel{\text{def}}{=} \forall t, n.\texttt{T}(t) \otimes \texttt{C}(n, breast, 1, [\texttt{EPCAM}]) \multimap \texttt{T}(t + d_{20}(1)) \otimes \texttt{C}(n, breast, 0, [\texttt{EPCAM}])$$

This rule describes a cell acquiring passenger mutations. Its fitness is decreased by one in a time-delay $d_{20}(1)$ and its driver mutations remain unchanged. In this formula, we have introduced two new connectives: the universal quantifier allowing us to instantiate the same rule for any cell n and any time-unit t and the multiplicative conjunction $\otimes$ whose rules are:

$$\frac{\Gamma; \Delta_1 \vdash \Downarrow F_1 \quad \Gamma; \Delta_2 \vdash \Downarrow F_2}{\Gamma; \Delta_1, \Delta_2 \vdash \Downarrow F_1 \otimes F_2} \otimes_R \qquad \frac{\Gamma; \Delta, F_1, F_2 \vdash G}{\Gamma; \Delta, F_1 \otimes F_2 \vdash G} \otimes_L$$

The rule $\otimes_R$ belongs to the positive phase and it says that the proof of $F_1 \otimes F_2$ requires to split the linear context in order to prove both F_1 and F_2. The left rule belongs to the negative phase and the resource $F_1 \otimes F_2$ is simply transformed into two resources (F_1 and F_2).

Most of the rules in our model are parametric on the fitness degree. Hence, a rule of the form:

$$\texttt{rl}(br_{t1}) \stackrel{\text{def}}{=} \forall t, n. \, \texttt{T}(t) \otimes \texttt{C}(n, breast, f, [\texttt{TGF}\beta]) \multimap \texttt{T}(t + d_{11}(f)) \otimes \texttt{A}(n), \, f \in 0..2$$

represents, in fact, three rules (one for each value of $f \in 0..2$). This family of rules describes three cases of apoptosis. In this particular example, any cell located in the breast, with fitness degree 0, 1, or 2 and list of mutations $[\texttt{TGF}\beta]$ may go to apoptosis and the time needed for such a transition is $d_{11}(f)$. Note that $d(\cdot)$ is a function that depends on f. If such $d(\cdot)$ does not depend on f, we shall simply write d instead of $d()$.

A typical rule describing a cell acquiring a driver mutation is:

$$\texttt{rl}(br_{t2.1}) \stackrel{\text{def}}{=} \forall t, n. \, \texttt{T}(t) \otimes \texttt{C}(n, breast, 1, [\texttt{TGF}\beta]) \multimap \texttt{T}(t + d_{12}) \otimes \texttt{C}(n, breast, 2, [\texttt{EPCAM}])$$

This rule says that a cell in the breast with a fitness degree $f = 1$ may acquire a new mutation ($\texttt{EPCAM}$), which increases its fitness by 1.

Another kind of rule describes a cell moving from one compartment to the next. The following rule describes an intravasating CTC:

$$\texttt{rl}(br_{e2}) \stackrel{\text{def}}{=} \forall t, n. \, \texttt{T}(t) \otimes \texttt{C}(n, breast, f, [\texttt{EPCAM}]) \multimap \texttt{T}(t + d_{22}(f)) \otimes \texttt{C}(n, blood, 2, [\texttt{EPCAM}]), \, f \in 1..3$$

Finally, a last kind of rule describes a DNA repair of passenger mutations:

$$\texttt{rl}(br_{e0r}) \stackrel{\text{def}}{=} \forall t, n.\texttt{T}(t) \otimes \texttt{C}(n, breast, f, [\texttt{EPCAM}]) \multimap \texttt{T}(t + d_{20r}(f)) \otimes \texttt{C}(n, breast, f + 1, [\texttt{EPCAM}]), \, f \in 1..2$$

The complete set of rules is in Fig. 1. For the sake of readability, we omit the universal quantification on t and n in the formulas. We shall use system to denote the set of rules and then, sequents take the form:

$$\text{system} ; \; T(t), C(\cdot), \cdots, C(\cdot) \vdash G$$

where G is a property to be proved (Sect. 4). Let s be a multiset of the form $\{C_1(n_1, c_1, f_1, lm_1), ..., C_n(n_n, c_n, f_n, lm_n), A(n_1'), ..., A(n_k')\}$ representing the state of different cells. The multiset $\{T(t), C_1(n_1, c_1, f_1, lm_1), ..., C_n(n_n, c_n, f_n, lm_n), A(n_1'), ..., A(n_k')\}$ of atomic formulas, formalising the state of the system at time-unit t, is denoted as $[\![s]\!]_t$. Observe that a cell that has gone to apoptosis cannot evolve any further (as no rule has a $A(n)$ on the left hand side).

In the breast
$r1(br_0)$ $\overset{\text{def}}{=}$ $T(t) \otimes C(n, breast, 1, [\,]) \multimap T(t + d_{00}) \otimes C(n, breast, 0, [\,])$
$r1(br_1)$ $\overset{\text{def}}{=}$ $T(t) \otimes C(n, breast, f, [\,]) \multimap T(t + d_{01}(f)) \otimes A(n) \; f \in 0 .. 1$
$r1(br_2)$ $\overset{\text{def}}{=}$ $T(t) \otimes C(n, breast, 1, [\,]) \multimap T(t + d_{02}) \otimes C(n, breast, 1, [\text{TGF}\beta])$
$r1(br_{t0})$ $\overset{\text{def}}{=}$ $T(t) \otimes C(n, breast, 1, [\text{TGF}\beta]) \multimap T(t + d_{10}) \otimes C(n, breast, 0, [\text{TGF}\beta])$
$r1(br_{t0r})$ $\overset{\text{def}}{=}$ $T(t) \otimes C(n, breast, 1, [\text{TGF}\beta]) \multimap T(t + d_{10r}) \otimes C(n, breast, 2, [\text{TGF}\beta])$
$r1(br_{t1})$ $\overset{\text{def}}{=}$ $T(t) \otimes C(n, breast, f, [\text{TGF}\beta]) \multimap T(t + d_{11}(f)) \otimes A(n), \; f \in 0 .. 2$
$r1(br_{t2})$ $\overset{\text{def}}{=}$ $T(t) \otimes C(n, breast, f, [\text{TGF}\beta]) \multimap T(t + d_{12}) \otimes C(n, breast, f + 1, [\text{EPCAM}]) \; f \in 1 .. 2$
$r1(br_{e0})$ $\overset{\text{def}}{=}$ $T(t) \otimes C(n, breast, f, [\text{EPCAM}]) \multimap T(t + d_{20}(f)) \otimes C(n, breast, f - 1, [\text{EPCAM}]), \; f \in 1 .. 3$
$r1(br_{e0r})$ $\overset{\text{def}}{=}$ $T(t) \otimes C(n, breast, f, [\text{EPCAM}]) \multimap T(t + d_{20r}(f)) \otimes C(n, breast, f + 1, [\text{EPCAM}]), \; f \in 1 .. 2$
$r1(br_{e1})$ $\overset{\text{def}}{=}$ $T(t) \otimes C(n, breast, f, [\text{EPCAM}]) \multimap T(t + d_{21}(f)) \otimes A(n), \; f \in 0 .. 3$
$r1(br_{e2})$ $\overset{\text{def}}{=}$ $T(t) \otimes C(n, breast, f, [\text{EPCAM}]) \multimap T(t + d_{22}(f)) \otimes C(n, blood, f + 1, [\text{EPCAM}]), \; f \in 1 .. 3$
In the blood
$r1(bl_{e0})$ $\overset{\text{def}}{=}$ $T(t) \otimes C(n, blood, f, [\text{EPCAM}]) \multimap T(t + d_{30}(f)) \otimes C(n, blood, f - 1, [\text{EPCAM}]), \; f \in 1 .. 4$
$r1(bl_{e0r})$ $\overset{\text{def}}{=}$ $T(t) \otimes C(n, blood, f, [\text{EPCAM}]) \multimap T(t + d_{30r}(f)) \otimes C(n, blood, f + 1, [\text{EPCAM}]), \; f \in 1 .. 3$
$r1(bl_{e1})$ $\overset{\text{def}}{=}$ $T(t) \otimes C(n, blood, f, [\text{EPCAM}]) \multimap T(t + d_{31}(f)) \otimes A(n), \; f \in 0 .. 4$
$r1(bl_{e2})$ $\overset{\text{def}}{=}$ $T(t) \otimes C(n, blood, f, [\text{EPCAM}]) \multimap T(t + d_{32}(f)) \otimes C(n, blood, f + 2, [\text{EPCAM}, \text{CD47}]), \; f \in 1 .. 4$
$r1(bl_{ec0})$ $\overset{\text{def}}{=}$ $T(t) \otimes C(n, blood, f, [\text{EPCAM}, \text{CD47}]) \multimap T(t + d_{40}(f)) \otimes C(n, blood, f - 1, [\text{EPCAM}, \text{CD47}]), \; f \in 1 .. 6$
$r1(bl_{ec1})$ $\overset{\text{def}}{=}$ $T(t) \otimes C(n, blood, f, [\text{EPCAM}, \text{CD47}]) \multimap T(t + d_{41}(f)) \otimes A(n), \; f \in 0 .. 6$
$r1(bl_{ec2})$ $\overset{\text{def}}{=}$ $T(t) \otimes C(n, blood, f, [\text{EPCAM}, \text{CD47}]) \multimap T(t + d_{42}(f)) \otimes C(n, blood, f + 2, [\text{EPCDCD}]), \; f \in 1 .. 6$
$r1(bl_{ec3})$ $\overset{\text{def}}{=}$ $T(t) \otimes C(n, blood, f, [\text{EPCAM}, \text{CD47}]) \multimap T(t + d_{43}(f)) \otimes C(n, blood, f + 2, [\text{EPCDME}]), \; f \in 1 .. 6$
$r1(bl_{ecc0})$ $\overset{\text{def}}{=}$ $T(t) \otimes C(n, blood, f, [\text{EPCDCD}]) \multimap T(t + d_{50}(f)) \otimes C(n, blood, f - 1, [\text{EPCDCD}]), \; f \in 1 .. 6$
$r1(bl_{ecc1})$ $\overset{\text{def}}{=}$ $T(t) \otimes C(n, blood, f, [\text{EPCDCD}]) \multimap T(t + d_{51}(f)) \otimes A(n), \; f \in 0 .. 8$
$r1(bl_{ecc2})$ $\overset{\text{def}}{=}$ $T(t) \otimes C(n, blood, f, [\text{EPCDCD}]) \multimap T(t + d_{52}(f)) \otimes C(n, blood, f + 2, [\text{EPCDCDME}]), \; f \in 1 .. 8$
$r1(bl_{ecm0})$ $\overset{\text{def}}{=}$ $T(t) \otimes C(n, blood, f, [\text{EPCDME}]) \multimap T(t + d_{60}(f)) \otimes C(n, blood, f - 1, [\text{EPCDME}]), \; f \in 1 .. 8$
$r1(bl_{ecm1})$ $\overset{\text{def}}{=}$ $T(t) \otimes C(n, blood, f, [\text{EPCDME}]) \multimap T(t + d_{61}(f)) \otimes A(n), \; f \in 0 .. 8$
$r1(bl_{ecm2})$ $\overset{\text{def}}{=}$ $T(t) \otimes C(n, blood, f, [\text{EPCDME}]) \multimap T(t + d_{62}(f)) \otimes C(n, blood, f + 2, [\text{EPCDCDME}]), \; f \in 1 .. 8$
$r1(bl_{eccm0})$ $\overset{\text{def}}{=}$ $T(t) \otimes C(n, blood, f, [\text{EPCDCDME}]) \multimap T(t + d_{70}(f)) \otimes C(n, blood, f - 1, [\text{EPCDCDME}]), \; f \in 1 .. 10$
$r1(bl_{eccm1})$ $\overset{\text{def}}{=}$ $T(t) \otimes C(n, blood, f, [\text{EPCDCDME}]) \multimap T(t + d_{71}(f)) \otimes A(n), \; f \in 0 .. 10$
$r1(bl_{eccm2})$ $\overset{\text{def}}{=}$ $T(t) \otimes C(n, blood, f, [\text{EPCDCDME}]) \multimap T(t + d_{72}) \otimes C(n, bone, f + 1, [\text{EPCDCDME}]), \; f \in 1 .. 10$
In the bone
$r1(bo_0)$ $\overset{\text{def}}{=}$ $T(t) \otimes C(n, bone, f, [\text{EPCDCDME}]) \multimap T(t + d_{80}(f)) \otimes C(n, bone, f - 1, [\text{EPCDCDME}]), \; f \in 1 .. 11$
$r1(bo_1)$ $\overset{\text{def}}{=}$ $T(t) \otimes C(n, bone, f, [\text{EPCDCDME}]) \multimap T(t + d_{81}(f)) \otimes A(n), \; f \in 0 .. 11$
$r1(bo_2)$ $\overset{\text{def}}{=}$ $T(t) \otimes C(n, bone, f, [\text{EPCDCDME}]) \multimap T(t + d_{82}(f)) \otimes C(n, bone, f + 1, [\text{EPCDCDME}, \text{seeded}]), \; f \in 1 .. 11$

Fig. 1. Complete set of rules. Variables t and n are universally quantified. EPCDCDME, EPCDCD and EPCDME are shorthand, respectively, for the list of mutations [EPCAM, CD47, CD44, MET], [EPCAM, CD47, CD44] and [EPCAM, CD47, MET].

We note that our rules are asynchronous: only one rule can be fired at a time. As in Biocham, we choose an asynchronous semantics in order to eliminate the risk of affecting fundamental biological phenomena such as the masking of a relation by another one and the consequent inhibition/activation of biological processes. Finally, we note that the delays depend on the fitness and more accurate values can be found using data. DNA mutational processes have been succesfully modeled as compound poisson processes (see for instance [16]). Delays could be seen as events waiting times which are well-known measures for Poisson processes (see for instance [25]). In our model, delays are (uninterpreted) logical constants that can be later tuned when experimental results are available. The proofs presented here remain the same regardless such values.

3.1   Adequacy

In this section, we prove the adequacy of our encoding: a single transition in the state of the system corresponds, exactly, to a complete focused phase (a positive phase followed by a negative phase) in focused LL [2]. Focusing organises proofs in phases where a negative phase introduces all the invertible rules. In a positive phase, we choose one of the formulas (notation $\Downarrow F$) whose principal connective has a non-invertible rule. Introducing such a connective is a *decision* in the proof search procedure and then, the order in which we apply them may lead to a proof or not. Proofs must finish in the positive phase with an initial rule:

$$\overline{\Gamma; p \vdash\Downarrow p}\ I \qquad \overline{\Gamma, p; \cdot \vdash\Downarrow p}\ I$$

where p is an atomic formula (e.g, $C_1(n, c, f, lm)$). Note that the proof ends when p is the only atom in the linear context or when the linear context is empty and p is in the classical context. Our results are based on the following observations:

1. It is not possible to focus on the formulas resulting from $[\![s]\!]_t$ since those are atoms. Hence, no focus step can start by focusing on those formulas.
2. Once we focus on one of the formulas in the classical context **system** (modelling the rules of the system), what we observe is that one of the $C(n, c, f, lm)$ formulas is consumed as well as the predicate $T(t)$. The focus phase ends by producing the needed $C(n, c', f', lm')$ and $T(t')$ atoms (or $A(n)$ in the case of apoptosis rules). This means that focused derivations are in one-to-one correspondence with steps in the system.

In the following theorem, we shall use the notation $s \xrightarrow{(r,d)} s'$ to denote that the system may evolve from state s to state s' by applying the rule r that takes d time-units. Hence, $S_s = \{(s', r, d) \mid s \xrightarrow{(r,d)} s'\}$ represents the set of possible transitions starting from s. We shall show that all transitions in S_s match exactly one focused derivation of the encoded system. More precisely (proof in Appendix)

Theorem 1 (Adequacy). *Let s be a state and $S_s = \{(s', r, d) \mid s \xrightarrow{(r,d)} s'\}$. Then, $(s', r, d) \in S_s$ iff focusing on the encoding of r leads to the following derivation.*

$$\frac{\texttt{system} \; ; [\![s']\!]_{t+d} \vdash G}{\texttt{system} \; ; [\![s]\!]_t \vdash G}$$

The following corollaries are immediate consequences of Theorem 1.

Corollary 1 (Adequacy). *Let s and s' be two states. Then $s \xrightarrow{(r,d)} s'$ iff the sequent $\texttt{system}; \cdot \vdash [\![s]\!]_t \multimap [\![s]\!]_{t+d}$ is provable.*

The above results allow us to use the whole positive-negative phase as macro rules in the logical system. Hence, during proofs, we shall abuse notation and we shall use, e.g., $\texttt{rl}(bo_2)$ as a logical inference rule. Formally, we can show that the corresponding logical rule is admissible in the system, i.e., if the premise is provable then the conclusion is also provable.

Corollary 2 (Macro rules). *Assume that $s \xrightarrow{(r,d)} s'$. Then, the following macro rule is admissible:*

$$\frac{\texttt{system} \; ; \Delta, [\![s']\!]_{t+d} \vdash G}{\texttt{system} \; ; \Delta, [\![s]\!]_t \vdash G} \; r$$

4   Verifying Properties of the Model

The goal of this section is twofold: testing our rules, but also testing some hypothesis of our model—as these are recent proposals in the literature. We shall detail some of the proofs here. The others can be found in the proof scripts and the documentation of our Coq formalisation (http://subsell.logic.at/bio-CTC/).

4.1   Reachability and Existence of Cycle Properties

Recall that a *Circulating Tumour Cell (CTC)* is a cancer cell in the blood, i.e., a cell $\texttt{C}(n, blood, f, m)$. An *extravasating CTC* is a CTC that has reached the bone, i.e., a cell of the shape $\texttt{C}(n, bone, f, [\texttt{EPCAM}, \texttt{CD47}, \texttt{CD44}, \texttt{MET}])$. A first property of interest might be the following one: "is it possible for a CTC, with mutations $[\texttt{EPCAM}, \texttt{CD47}]$ and fitness 3, to become an extravasating CTC with fitness 8? What is the time delay for such a transition?" This is formalised as follows:

Property 1. The following sequent is provable:

$$\texttt{system} \; ; \; . \vdash \forall n, t. \; \texttt{T}(t) \otimes \texttt{C}(n, blood, 3, [\texttt{EPCAM}, \texttt{CD47}])$$
$$\multimap \exists d. \; \texttt{T}(t + d) \otimes \texttt{C}(n, bone, 8, [\texttt{EPCAM}, \texttt{CD47}, \texttt{CD44}, \texttt{MET}])$$

In our Coq formalisation, we have implemented several tactics (e.g., `solveF` and `applyRule` used below) to automate the process of proving properties and make the resulting scripts compact and clear. This should ease the testing/proving of new hypotheses in our model. For instance, the proof of the previous property is as follows (F below is the formula in Property 1):

```
Lemma Property1: forall n t, |- System ; F
Proof with solveF . (* solves the "trivial" goals in a focused proof *)
  intros. (* introducing the quantified variables n and t *)
  applyRule (blec2 3). (* application of macro rules -- corollary 2-- *)
  applyRule (blecc2 5).
  applyRule (bleccm2 7).
  eapply tri_dec1 ... (* decision rule, focusing on the goal *)
  eapply tri_tensor ... (* tensor *)
  eapply tri_ex with (t:= (d72 7) s+ (d52 5) s+ (d42 3) s+ (Cte t)) ... (* existential quantif. *)
  eapply Init1... (* initial rule *)
  eapply Init1... (* initial rule *)
Qed.
```

The reader may compare the steps in the script above with the proof (by hand) of Property 1 in Appendix B.

Our next property is the following: "what is the time delay for a CTC with mutations [EPCAM, CD47] and fitness 3 or 4 to become an extravasating CTC with fitness between 6 and 9?"

Property 2. The following sequent is provable:

$$\text{system} \; ; \; \vdash \forall n, t. \; \mathsf{T}(t) \otimes (\mathsf{C}(n, blood, 3, [\text{EPCAM}, \text{CD47}]) \oplus \mathsf{C}(n, blood, 4, [\text{EPCAM}, \text{CD47}])) \multimap$$
$$\exists t_d. \; \mathsf{T}(t + t_d) \otimes$$
$$(\mathsf{C}(n, bone, 6, [\text{EPCAM}, \text{CD47}, \text{CD44}, \text{MET}]) \oplus \mathsf{C}(n, bone, 7, [\text{EPCAM}, \text{CD47}, \text{CD44}, \text{MET}]) \oplus$$
$$\mathsf{C}(n, bone, 8, [\text{EPCAM}, \text{CD47}, \text{CD44}, \text{MET}]) \oplus \mathsf{C}(n, bone, 9, [\text{EPCAM}, \text{CD47}, \text{CD44}, \text{MET}]))$$

Due to the $\oplus$ connective (additive conjunction), the proof of this property entails two proof obligations (see the details in Appendix B): the case when the fitness is 3 and the case when the fitness is 4. The proof of the first case reveals that the rules $\mathtt{rl}(bl_{ec0})$ and $\mathtt{rl}(bl_{ecc0})$ could be used zero or more times—as long as the fitness remains positive. In the end, when the state $\mathsf{C}(n, bone, 6, [\text{EPCAM}, \text{CD47}, \text{CD44}, \text{MET}])$ is reached, we also obtain in our proof the delay needed ($t_d = d_{40}(3) + d_{42}(2) + d_{50}(4) + d_{52}(3) + d_{72}(5)$) to reach that state. The second proof obligation can be discharged by considering several paths, depending on the order of mutations CD44 and MET and the eventually many passenger mutations (rules $bl_{ec0}, bl_{ecc0}, bl_{ecc0}$, and bl_{ecm0}). We illustrate some of those paths in Appendix B. Finally, we note that along with the time delay t_d we are looking for, the proof provides also the fitness of the extravasating CTC.

Existence of Cycle. Rules for passenger mutations decrease the fitness of the cell by one, while rules for DNA repair increase the fitness. Hence, we may observe loops and oscillations in our model. This can be exemplified in the following property: "a cell in the breast, with mutation [EPCAM], might have its fitness oscillating from 1 to 2 and back."

Property 3. The following sequents are provable:

$$\text{system} \; ; \; . \vdash \forall n, t. \; \mathsf{T}(t) \otimes \mathsf{C}(n, breast, 1, [\text{EPCAM}]) \multimap \exists d. \; \mathsf{T}(t + d) \otimes \mathsf{C}(n, breast, 2, [\text{EPCAM}]) \; \text{and}$$
$$\text{system} \; ; \; . \vdash \forall n, t. \; \mathsf{T}(t) \otimes \mathsf{C}(n, breast, 2, [\text{EPCAM}]) \multimap \exists d. \; \mathsf{T}(t + d) \otimes \mathsf{C}(n, breast, 1, [\text{EPCAM}])$$

4.2 Meta-Level Properties

In a first experiment on using LL for biology on the computer [12], we defined the set of biological rules as an inductive type in Coq, and proved some of their properties by induction on the set of fireable rules. Here, we choose a different approach. We have defined the biological rules by formulas in LL, and we use focusing, along with adequacy, to look for the fireable rules at a given state. Properties whose proofs need meta-reasoning will be formalised at the level of derivations. In this section, we give two examples of these.

Let us first give a simple property concerning apoptosis, namely: "any cell having a null fitness must go to apoptosis".

Property 4. Let Δ be a multiset of atoms of the form $\mathtt{C}(\cdot)$. Then, in any derivation of the form

$$\frac{\mathtt{system};\Delta,\mathtt{T}(t+d),St \vdash G}{\mathtt{system};\Delta,\mathtt{T}(t),\mathtt{C}(n,c,0,m) \vdash G}\ \mathtt{rl}(\cdot)$$

we have $St = \mathtt{A}(n)$.

In our Coq formalisation, the above property can be discharged with few lines of code:

```
Lemma Property4: forall n t c lm , F.
Proof by solveF .
 intros H. (* the first sequent is assumed to be provable *)
 apply FocusOnlyTheory in H;auto. (* The proof H must start by focusing on one of the rules *)
 destruct H as [R] ;destruct H.
 repeat first [ CaseRule | DecomposeRule; FindUnification | eauto ].
Qed.
```

The `FocusOnlyTheory` lemma says that the proof of the sequent must start by focusing on one of the formulas in `System`. The `destruct` tactic simplifies the hypotheses after the use of lemma `FocusOnlyTheory`. The interesting part is the last line of the script. The `CaseRule` tactic tests each of the rules of the system. Then, `DecomposeRule; FindUnification` decomposes (positive-negative phase) the application of the rule. Finally, `eauto` proves the desired goal after the application of the rule. This is a very general scheme, where we do case analysis on all possible rules. Some of them cannot be fired in the current state and then, the proof follows by contradiction. In the rest of the cases, the `eauto` tactic is able to conclude the goal.

The following property states one of the key properties of our model: "any cell in the blood, with mutations including `CD47`, has four possible evolutions:

1. acquiring passenger mutations: its fitness decreases by one and the driver mutations remain unchanged;
2. going to apoptosis;
3. acquiring a driver mutation: its fitness increases by two;
4. moving to the bone: its fitness increases by one and the driver mutations ([`EPCAM`, `CD47`, `CD44`, `MET`]) remain unchanged."

Property 5. Let Δ be a multiset of atoms of the form $\mathtt{C}(\cdot)$. Then, in any derivation of the form

$$\frac{\mathtt{system};\, \Delta, \mathtt{T}(t + t_d), St \vdash G}{\mathtt{system};\, \Delta, \mathtt{T}(t), \mathtt{C}(n, blood, f, m) \vdash G} \ \mathtt{rl}(\cdot)$$

with m containing CD47, it must be the case that

1. either $St = \mathtt{C}(n, blood, f - 1, m)$,
2. or $St = \mathtt{A}(n)$,
3. or $St = \mathtt{C}(n, blood, f + 2, m')$ with m' being as m plus an additional mutation,
4. or $St = \mathtt{C}(n, bone, f + 1, m)$ with $m = [\text{EPCAM}, \text{CD47}, \text{CD44}, \text{MET}]$.

The proof follows the same rationale as in Property 4. In fact, the proof in Coq is exactly the same but, in the last line, we have

```
repeat (first [ CaseRule | DecomposeRule; FindUnification | SolveGoal]).
```

where `SolveGoal` (instead of `eauto`) is able to finish the resulting cases.

5 Concluding Remarks

Our goal is to study cancer progression, aiming at a better understanding of it, and, in the long term, help in finding, and testing, new targeted drugs, a priori much more efficient than most of the drugs used so far. This paper describes the use of linear logic in modelling the multi compartment role of driver mutations in breast cancer. This work is innovative but also proof-of-principle. It can clearly be generalised to other cancer types where driver mutations are known. It also makes evident the capability of this logical approach to integrate different types of data and output a diagnosis with higher interpretability than many currently fashionable machine learning methods such as deep learning. Note however that building a system for cancer/disease diagnosis and therapy prognosis would require both automatic proof search and taking the size of the tumour into account. Also note that, although all the properties considered so far only deal with the evolution of one single cell, our approach allows us to consider a state with many cells. We believe that the paper and the rich sections in the online supplementary material would become an important resource for other similar studies.

For example we can believe there is a complementary of our work with respect to mathematical models such as [4] that include survival data and quantitative results. Logic allows to model the evolution of cells across scales and compartments, while ODEs require parameter estimation (qualitative vs quantitative).

While temporal logics have been very successful in practice with efficient model checking tools, these logics do not enjoy standard proof theory. In contrast, LL has a very traditional proof theoretic pedigree: it is naturally presented as a sequent calculus that enjoys cut-elimination and focusing. A further advantage of our approach wrt model checking is that it provides a unified framework to encode both transition rules and (both statements and proofs of) temporal properties. Observe also that we do not need to build the set of states of the transition system. We view model checking as a useful first step before proofs: testing the model before trying to prove properties of it. The interested reader can find a detailed comparison of the approaches in [12]. See also [14] for an adequate encoding of Temporal Logic in LL.

In order to describe *constrained* transitions systems (by timed or spatial restrictions for example), previous work have proposed different extensions of Linear Logic, e.g., HyLL (a modal extension of ILL) [13] and SELL$^\cap$ (LL with with quantifiers on subexponentials). See [14] for a formal comparison of the two logics, and an adequate encoding of Temporal Logic in LL extended with fixpoints. These Logical Frameworks were then used to specify and analyse small biology systems in two initial experiments: [12] in HyLL and [28] in SELL$^\cap$. In the present work, we chose to use pure LL, and define a predicate to encode time. This approach avoids the extra complication of copying the "unused" information to the next time-unit (HyLL world or SELL subexponential), thus benefiting from the usual compositional nature of the logic.

Future Perspectives. The ongoing revolution in AI is accelerating the development of software that enables computers to perform "intelligent" clinical and medical tasks. Machine learning algorithms find hidden patterns in data, classify and associate similar patients/diseases/drugs based on common features (e.g., the IBM Watson system which is used to analyse genomic and cancer data). Future challenges in medicine include understanding bias in data collection (and also in doctor's experience) and fostering the ability to integrate evidence from heterogeneous datasets, from different omics and clinical data, from several lines of independent data. We believe that machine learning could satisfy well these needs and that there is also a need to develop methods that offer a hypothesis-driven approach, so that doctors do not feel that they are going to be replaced. Such methods could provide them with a personalised and easily interpretable clinical support decision-making tool that could perform a synthesis of qualitative and quantitative multi-modal evidence. Examples of decision trees used in current practice for breast cancer diagnosis can be found at pages 598–603 of [27]. Our logical approach, although focused on driver mutations, goes in such a direction and could be used with continuous and discrete mixed variables. This information could be obtainable through single cell experiments on cancer biopsies (although with large variance), which is now at the stage of passing from basic science to clinical protocols. Machine learning could analyse cancer mutation patterns and feed our logic approach with this information that could be integrated with other rules such as changes on the metabolic networks or on epigenetics. Other rules could be derived from other levels of cancer clinical investigation such as from image data (changes in fMRI, CT-scans and microscopy samples), blood analyses (identification and counts of circulating cancer cells) and other types of medical observations. The long term plan is to build a portable resource that facilitates diagnostic and therapeutic decision making and promotes a cost-effective personalised patient workup. This would represent a new paradigm in personalised and precision cancer treatment which integrates multi-modality analyses and clinical characteristics in a near-real time manner, improving clinical management of cancer. Finally we believe that logical approaches could improve the harmonisation and standardisation of the reporting and interpretation of clinically relevant data.

Acknowledgements. We thank the anonymous reviewers for their valuable comments on an earlier draft of this paper. The work of Olarte was supported by CNPq, the project FWF START Y544-N23 and by CAPES, Colciencias, and INRIA via the STIC AmSud project EPIC (Proc. No 88881.117603/2016-01).

Appendix

A   Proof of the Adequacy Results

Theorem 1: Adequacy. Let s be a state and $S_s = \{(s', r, d) \mid s \xrightarrow{(r,d)} s'\}$. Then, $(s', r, d) \in S_s$ iff focusing on the encoding of r leads to the following derivation.

$$\frac{\text{system} \; ; [\![s']\!]_{t+d} \vdash G}{\text{system} \; ; [\![s]\!]_t \vdash G}$$

Proof. The encoding of the rule r is a bipole [2] (i.e., a formula that, being focused, will produce a single positive and a single negative phases) of the form

$$\forall t, n. \; \mathsf{T}(t) \otimes \mathsf{C} \multimap \mathsf{T}(t+d) \otimes \mathsf{C}'$$

Focusing on this formula (stored in system) necessarily produces the following derivation, starting with rule D_C (decision on the classical context):

$$\frac{\dfrac{\overset{\pi}{\text{system} ; \Delta \vdash \Downarrow \mathsf{T}(t) \otimes \mathsf{C}} \quad \overset{\psi}{\text{system} ; \Delta', \Downarrow \mathsf{T}(t+d) \otimes \mathsf{C}' \vdash G}}{\dfrac{\text{system} ; [\![s]\!]_t, \Downarrow \mathsf{T}(t) \otimes \mathsf{C} \multimap \mathsf{T}(t+d) \otimes \mathsf{C}' \vdash G}{\dfrac{\text{system} ; [\![s]\!]_t, \Downarrow \forall t, n. \; \mathsf{T}(t) \otimes \mathsf{C} \multimap \mathsf{T}(t+d) \otimes \mathsf{C}' \vdash G}{\text{system} ; [\![s]\!]_t \vdash G} \; \begin{matrix} \forall_L \times 2 \\ D_C \end{matrix}} \; \multimap_L}$$

Here (Δ, Δ') is a partition of the atoms in $[\![s]\!]_t$. Since r is fireable in the state s, then Δ must contain all the atoms needed to prove $\mathsf{T}(t)$ and C. Moreover, Δ' must correspond to the components not affected by the application of r, i.e., $\Delta' = [\![s]\!]_t \setminus \Delta$. Hence, derivation π takes the form:

$$\frac{\overline{\text{system} ; \mathsf{T}(t) \vdash \Downarrow \mathsf{T}(t)} \; I \quad \overline{\text{system} ; \mathsf{C} \vdash \Downarrow \mathsf{C}} \; I}{\text{system} ; \Delta \vdash \Downarrow \mathsf{T}(t) \otimes \mathsf{C}} \; \otimes_R$$

This means that $\Delta = \{\mathsf{T}(t), \mathsf{C}\}$. On the other hand, derivation ψ starts with the release rule R (since $\otimes_L$ must be introduced in the negative phase and then, focusing is lost) and we have

$$\frac{\text{system} ; \Delta', \mathsf{T}(t+d), \mathsf{C}' \vdash G}{\text{system} ; \Delta', \Downarrow \mathsf{T}(t+d) \otimes \mathsf{C}' \vdash G} \; R, \otimes_L$$

In the last sequent, the negative phase ends. Note that the set $\{\mathsf{T}(t+d), \mathsf{C}'\}$ corresponds to $[\![s']\!]_{t+d}$.

Corollary 1: Adequacy. Let s and s' be two states. Then $s \xrightarrow{(r,d)} s'$ iff the sequent $\texttt{system}; \cdot \vdash [\![s]\!]_t \multimap [\![s]\!]_{t+d}$ is provable.

Proof. Note that after the negative phase, we have:

$$\frac{\texttt{system}; \Delta \vdash [\![s]\!]_{t+d}}{\texttt{system}; \cdot \vdash [\![s]\!]_t \multimap [\![s]\!]_{t+d}}$$

where Δ is the multiset of atoms in $[\![s]\!]_t$. We cannot focus on those atoms (since they are positive). Moreover, we cannot focus on $[\![s]\!]_{t+d}$ (since the atom $\texttt{T}(t+d)$ is not in Δ nor in $\texttt{system}$). Hence, we can only focus on the formulas in $\texttt{system}$. We conclude by focusing on the encoding of r and using Theorem 1.

The proof of Corollary 2 follows easily from Theorem 1.

B   Proof of the Properties of the Model

Property 1. The following sequent is provable:

$$\texttt{system}\ ;\ .\vdash \forall n, t.\ \texttt{T}(t) \otimes \texttt{C}(n, blood, 3, [\texttt{EPCAM}, \texttt{CD47}]) \\ \multimap \exists d.\ \texttt{T}(t+d) \otimes \texttt{C}(n, bone, 8, [\texttt{EPCAM}, \texttt{CD47}, \texttt{CD44}, \texttt{MET}])$$

Proof. After the negative phase (using the rules $\forall_R$, $\otimes_L$, $\multimap_R$), we have only one proof obligation: $\texttt{system}\ ;\ \texttt{C}(n, blood, 3, [\texttt{EPCAM}, \texttt{CD47}]), \texttt{T}(t) \vdash G$, where G is

$$\exists t_d.\texttt{T}(t+t_d) \otimes \texttt{C}(n, bone, 8, [\texttt{EPCAM}, \texttt{CD47}, \texttt{CD44}, \texttt{MET}])$$

Note that the only non-atomic formulas are G and those formulas in $\texttt{system}$. The proof proceeds by focusing, several times, on the formulas in $\texttt{system}$ thus transforming the state $\texttt{C}(n, blood, 3, [\texttt{EPCAM}, \texttt{CD47}])$. In the end, we focus on G and the proof ends. Using the rules of the system as macro logical rules (see Corollary 2), we have the following:

$$\cfrac{\cfrac{\cfrac{\overset{\textstyle \pi}{\texttt{system}\ ;\ \texttt{C}(n, bone, 8, [\texttt{EPCDCDME}]), \texttt{T}(t + d_{42}(3) + d_{52}(5) + d_{72}(7)) \vdash G}}{\texttt{system}\ ;\ \texttt{C}(n, blood, 7, [\texttt{EPCDCDME}]), \texttt{T}(t + d_{42}(3) + d_{52}(5)) \vdash G}\ \texttt{rl}(bl_{eccm2.7})}{\texttt{system}\ ;\ \texttt{C}(n, blood, 5, [\texttt{EPCAM}, \texttt{CD47}, \texttt{CD44}]), \texttt{T}(t + d_{42}(3)) \vdash G}\ \texttt{rl}(bl_{ecc2.5})}{\texttt{system}\ ;\ \texttt{C}(n, blood, 3, [\texttt{EPCAM}, \texttt{CD47}]), \texttt{T}(t) \vdash G}\ \texttt{rl}(bl_{ec2.3})$$

In the above derivation, we note that, in the last sequent (bottom-up) we already reach the state $\texttt{C}(n, bone, 8, [\texttt{EPCAM}, \texttt{CD47}, \texttt{CD44}, \texttt{MET}])$, with delay $t_d = d_{42}(3) + d_{52}(5) + d_{72}(7)$. Hence, the derivation π corresponds to focusing and decomposing entirely the formula G:

$$\cfrac{\cfrac{\overline{\texttt{system}\ ;\texttt{T}(t + t_d) \vdash\Downarrow \texttt{T}(t + t_d)}\ ^I \quad \overline{\texttt{system}\ ;\texttt{C}(n, bone, 8, [\texttt{EPCDCDME}]) \vdash\Downarrow \texttt{C}(n, bone, 8, [\texttt{EPCDCDME}])}\ ^I}{\texttt{system}\ ;\cdots \vdash\Downarrow \texttt{T}(t + t_d) \otimes \texttt{C}(n, bone, 8, [\texttt{EPCDCDME}])}\ \otimes_R}{\texttt{system}\ ;\texttt{C}(n, bone, 8, [\texttt{EPCDCDME}]), \texttt{T}(t + t_d) \vdash\Downarrow G}\ \exists_R$$

Property 2. The following sequent is provable:

system ; $\vdash \forall n, t.\ \mathsf{T}(t) \otimes (\mathsf{C}(n, blood, 3, [\text{EPCAM}, \text{CD47}]) \oplus \mathsf{C}(n, blood, 4, [\text{EPCAM}, \text{CD47}])) \multimap$
$$\exists t_d.\ \mathsf{T}(t + t_d) \otimes$$
$$(\mathsf{C}(n, bone, 6, [\text{EPCAM}, \text{CD47}, \text{CD44}, \text{MET}]) \oplus \mathsf{C}(n, bone, 7, [\text{EPCAM}, \text{CD47}, \text{CD44}, \text{MET}]) \oplus$$
$$\mathsf{C}(n, bone, 8, [\text{EPCAM}, \text{CD47}, \text{CD44}, \text{MET}]) \oplus \mathsf{C}(n, bone, 9, [\text{EPCAM}, \text{CD47}, \text{CD44}, \text{MET}]))$$

Proof. After the negative phase (using the rules $\forall_R$, $\otimes_L$, $\oplus_L$, $\multimap_R$), we have two proof obligations (due to $\oplus_L$)

$$\text{(PO1)}\quad \text{system} \ ; \mathsf{C}(n, blood, 3, [\text{EPCAM}, \text{CD47}]), \mathsf{T}(t) \vdash G$$
$$\text{(PO2)}\quad \text{system} \ ; \mathsf{C}(n, blood, 4, [\text{EPCAM}, \text{CD47}]), \mathsf{T}(t) \vdash G$$

where G is the goal

$\exists t_d. \mathsf{T}(t + t_d) \otimes (\ \mathsf{C}(n, bone, 6, [\text{EPCAM}, \text{CD47}, \text{CD44}, \text{MET}]) \oplus$	first choice
$\mathsf{C}(n, bone, 7, [\text{EPCAM}, \text{CD47}, \text{CD44}, \text{MET}]) \oplus$	second choice
$\mathsf{C}(n, bone, 8, [\text{EPCAM}, \text{CD47}, \text{CD44}, \text{MET}]) \oplus$	third choice
$\mathsf{C}(n, bone, 9, [\text{EPCAM}, \text{CD47}, \text{CD44}, \text{MET}]))$	last choice

Let us start with the proof obligation (PO1). Similar to the proof of Property 1, we start by focusing on the formulas in system so that we may later focus on G. One of the possible paths/proofs leading to the conclusion of the goal G is the following:

$$\cfrac{\cfrac{\cfrac{\cfrac{\cfrac{\overset{\pi}{\text{system} \ ; \mathsf{C}(n, bone, 6, [\text{EPCDCDME}]), \mathsf{T}(t + d_{40}(3) + d_{42}(2) + d_{50}(4) + d_{52}(3) + d_{72}(5)) \vdash G}}{\text{system} \ ; \mathsf{C}(n, blood, 5, [\text{EPCDCDME}]), \mathsf{T}(t + d_{40}(3) + d_{42}(2) + d_{50}(4) + d_{52}(3)) \vdash G}\ \mathtt{rl}(bl_{eccm2.5})}{\text{system} \ ; \mathsf{C}(n, blood, 3, [\text{EPCDCD}]), \mathsf{T}(t + d_{40}(3) + d_{42}(2) + d_{50}(4)) \vdash G}\ \mathtt{rl}(bl_{ecc2.3})}{\text{system} \ ; \mathsf{C}(n, blood, 4, [\text{EPCDCD}]), \mathsf{T}(t + d_{40}(3) + d_{42}(2)) \vdash G}\ \mathtt{rl}(bl_{ecc0.4})}{\text{system} \ ; \mathsf{C}(n, blood, 2, [\text{EPCAM}, \text{CD47}]), \mathsf{T}(t + d_{40}(3)) \vdash G}\ \mathtt{rl}(bl_{ec2.2})}{\text{system} \ ; \mathsf{C}(n, blood, 3, [\text{EPCAM}, \text{CD47}]), \mathsf{T}(t) \vdash G}\ \mathtt{rl}(bl_{ec0.3})$$

In such a derivation, the rules $\mathtt{rl}(bl_{ec0})$ and $\mathtt{rl}(bl_{ecc0})$ could be used zero or more times - as long as the fitness remains positive. Moreover, in the last sequent (bottom-up) we have already reached the state $\mathsf{C}(n, bone, 6, [\text{EPCAM}, \text{CD47}, \text{CD44}, \text{MET}])$, with delay $t_d = d_{40}(3) + d_{42}(2) + d_{50}(4) + d_{52}(3) + d_{72}(5)$ and derivation π proceeds as in the proof of Property 1.

The proof obligation (PO2) can be discharged similarly by several paths, depending (as in PO1) on the order of mutations CD44 and MET and the eventually many passenger mutations (rules bl_{ec0}, bl_{ecc0}, bl_{ecc0}, and bl_{ecm0}). We give here the shortest path and one of the longest paths, as an illustration.

$\mathsf{C}(n, blood, 4, [\text{EPCAM}, \text{CD47}]) \otimes \mathsf{T}(t)$
$\multimap \mathsf{T}(t + d_{42}(4)) \otimes \mathsf{C}(n, blood, 6, [\text{EPCAM}, \text{CD47}, \text{CD44}]) - \mathtt{rl}(bl_{ec2.4})$
$\multimap \mathsf{T}(t + d_{42}(4) + d_{52}(6)) \otimes \mathsf{C}(n, blood, 8, [\text{EPCAM}, \text{CD47}, \text{CD44}, \text{MET}]) - \mathtt{rl}(bl_{ecc2.6})$
$\multimap \mathsf{T}(t + d_{42}(4) + d_{52}(6) + d_{72}(8)) \otimes \mathsf{C}(n, bone, 9, [\text{EPCAM}, \text{CD47}, \text{CD44}, \text{MET}]) - \mathtt{rl}(bl_{eccm2.8})$

$\mathsf{C}(n, blood, 4, [\text{EPCAM}, \text{CD47}]) \otimes \mathsf{T}(t)$
$\multimap \mathsf{T}(t + d_{40}(4)) \otimes \mathsf{C}(n, blood, 3, [\text{EPCAM}, \text{CD47}]) - \mathtt{rl}(bl_{ec0.4})$
$\multimap \mathsf{T}(t + d_{40}(4) + d_{43}(3)) \otimes \mathsf{C}(n, blood, 4, [\text{EPCAM}, \text{CD47}, \text{MET}]) - \mathtt{rl}(bl_{ec3.3})$
$\multimap \mathsf{T}(t + d_{40}(4) + d_{43}(3) + d_{60}(4)) \otimes \mathsf{C}(n, blood, 3, [\text{EPCAM}, \text{CD47}, \text{MET}]) - \mathtt{rl}(bl_{ecm0.4})$
$\multimap \mathsf{T}(t + d_{40}(4) + d_{43}(3) + d_{60}(4) + d_{62}(3)) \otimes \mathsf{C}(n, blood, 5, [\text{EPCAM}, \text{CD47}, \text{CD44}, \text{MET}]) - \mathtt{rl}(bl_{ecm2.3})$
$\multimap \mathsf{T}(t + d_{40}(4) + d_{43}(3) + d_{60}(4) + d_{62}(3) + d_{72}(5)) \otimes \mathsf{C}(n, bone, 6, [\text{EPCAM}, \text{CD47}, \text{CD44}, \text{MET}]) - \mathtt{rl}(bl_{ecm2.5})$

Note that along with the time delay t_d we are looking for, the proof provides also the fitness of the extravasating CTC.

Property 3. The following sequents are provable:

system ; . ⊢ ∀n, t. T(t) ⊗ C($n, breast$, 1, [EPCAM]) ⊸ ∃d. T($t + d$) ⊗ C($n, breast$, 2, [EPCAM]) and

system ; . ⊢ ∀n, t. T(t) ⊗ C($n, breast$, 2, [EPCAM]) ⊸ ∃d. T($t + d$) ⊗ C($n, breast$, 1, [EPCAM])

Proof. In this case we present the Coq script needed to discard this proof. We prove separately the two sequents above:

```
Lemma Property3_Seq1: forall n t,
    exists d,
|-F- Theory ; [ E{ fun _ x ⇒ perp TX{ fc1 d (var x)}} ** (C{ n; breast; 2; EP} ) ;
Atom T{ Cte t} ; Atom C{ n ; breast ; 1 ; EP}] ; UP [] .
Proof with solveF .
    idtac "Property3: Proving Cycle 1" .
    intros.
    eexists.
    applyRule (breOr 1).
    (* Proving the goal *)
    eapply tri_dec1
        with (F:= E{ fun (T : Type) (x : T) ⇒ perp TX{ DX{ d20 2, var x}}} ** C{ n; breast; 2; EP} ) ...
    eapply tri_tensor ...
    eapply tri_ex with (t:= (Cte t)) ...
    eapply Init1...
    eapply Init1...
Qed.

Lemma Property3_Seq2: forall n t,
    exists d,
|-F- Theory ; [ E{ fun _ x ⇒ perp TX{ fc1 d (var x)}} ** (C{ n; breast; 1; EP} ) ;
Atom T{ Cte t} ; Atom C{ n ; breast ; 2 ; EP}] ; UP [] .
Proof with solveF .
    idtac "Property3: Proving Cycle 2" .
    intros.
    eexists.
    applyRule (breO 2).
    (* Proving the goal *)
    eapply tri_dec1 with
        (F:= E{ fun (T : Type) (x : T) ⇒ perp TX{ DX{ d20 2, var x}}} ** C{ n; breast; 1; EP} ) ...
    eapply tri_tensor ...
    eapply tri_ex with (t:= (Cte t)) ...
    eapply Init1...
    eapply Init1...
Qed.
```

Property 4. Let Δ be a multiset of atoms of the form C($\cdot$). Then, in any derivation of the form

$$\frac{\text{system}; \Delta, \text{T}(t + d), St \vdash G}{\text{system}; \Delta, \text{T}(t), \text{C}(n, c, 0, m) \vdash G} \; \text{rl}(\cdot)$$

we have $St = \text{A}(n)$.

Proof. We know that the above derivation must start by focusing on one of the formulas in **system** (Theorem FocusOnlyTheory in our formalisation). Then, we proceed by case analysis on all of the rules. If the rule is not fireable, then we cannot focus on that rule since the initial rule cannot be applied (and the above derivation is not valid). If the rule can be fired, due to Corollary 2, we know that the resulting St is necessarily the one-step transformation of C($n, c, 0, m$), that, in this case, satisfies $St = \text{A}(n)$.

Property 5. Let Δ be a multiset of atoms of the form $\texttt{C}(\cdot)$. Then, in any derivation of the form

$$\frac{\texttt{system}; \Delta, \texttt{T}(t + t_d), St \vdash G}{\texttt{system}; \Delta, \texttt{T}(t), \texttt{C}(n, blood, f, m) \vdash G} \; \texttt{rl}(\cdot)$$

with m containing $\texttt{CD47}$, it must be the case that

1. either $St = \texttt{C}(n, blood, f - 1, m)$,
2. or $St = \texttt{A}(n)$,
3. or $St = \texttt{C}(n, blood, f + 2, m')$ with m' being as m plus an additional mutation,
4. or $St = \texttt{C}(n, bone, f + 1, m)$ with $m = [\texttt{EPCAM}, \texttt{CD47}, \texttt{CD44}, \texttt{MET}]$.

Proof. In this case we present the Coq script needed to discard this proof. Definition GoalP5 is just a shorthand to denote the goal we need to prove.

```
Definition GoalP5 (f n t m:nat) :=
 |-F- Theory ; [ Atom T{ Cte t} ; Atom C{ n ; blood; f ; m} ] ; UP [] →
                               In m [EPCDCD ;EPCDCDME;EPCDCDMEse] →
   exists (d m' :nat),
   |-F- Theory ; [ Atom T{ d s+ Cte t} ; Atom C{ n ;blood; f -1; m } ] ; UP [] ∨
   |-F- Theory ; [ Atom T{ d s+ Cte t} ; Atom A{ n} ] ; UP [] ∨
   (|-F- Theory ; [ Atom T{ d s+ Cte t} ; Atom C{ n ;blood; f +2; m' } ] ; UP [] ∧ (plusOne m m') ) ∨
   (m = EPCDCDME ∧ |-F- Theory ; [ Atom T{ d s+ Cte t} ; Atom C{ n ;bone; f + 1; EPCDCDME } ] ; UP []) .

Proposition Property5 : forall (f n t m:nat) ,
   GoalP5 f n t m.
 idtac "Proving Property 5".
 intros f n t m HProof HCaseM.
 apply FocusOnlyTheory in HProof;auto;
   destruct HProof as [R HProof]; destruct HProof as [HIn HProof];
    time "Solve:" repeat (first [ CaseRule | DecomposeRule; FindUnification | SolveGoal]) .
Qed.
```

References

1. Alur, R., et al.: Hybrid modeling and simulation of biomolecular networks. In: Di Benedetto, M.D., Sangiovanni-Vincentelli, A. (eds.) HSCC 2001. LNCS, vol. 2034, pp. 19–32. Springer, Heidelberg (2001). https://doi.org/10.1007/3-540-45351-2_6
2. Andreoli, J.: Logic programming with focusing proofs in linear logic. J. Log. Comput. **2**(3), 297–347 (1992)
3. Antczak, C., Mahida, J., Singh, C., Calder, P., Djaballah, H.: A high content assay to assess cellular fitness. Comb. Chem. High Throughput Screening **17**(1), 12–24 (2014)
4. Ascolani, G., Occhipinti, A., Lio, P.: Modeling circulating tumour cells for personalised survival prediction in metastatic breast cancer. PLoS Comput. Biol. **11**(5) (2015)
5. Bellomo, N., Preziosi, L.: Modelling and mathematical problems related to tumor evolution and its interaction with the immune system. Math. Comput. Model. **32**(3), 413–452 (2000)
6. Bertot, Y., Castéran, P.: Interactive Theorem Proving and Program Development. Coq'Art: The Calculus of Inductive Constructions. Springer, Heidelberg (2004). https://doi.org/10.1007/978-3-662-07964-5

7. Blinov, M.L., Faeder, J.R., Goldstein, B., Hlavacek, W.S.: BioNetGen: software for rule-based modeling of signal transduction based on the interactions of molecular domains. Bioinformatics **20**(17), 3289–3291 (2004)
8. Caravagna, G., et al.: Detecting repeated cancer evolution from multi-region tumor sequencing data. Nature Methods **15**(9), 707–714 (2018)
9. Chaouiya, C., Naldi, A., Remy, E., Thieffry, D.: Petri net representation of multi-valued logical regulatory graphs. Natural Comput. **10**(2), 727–750 (2011)
10. Clarke, E.M., Henzinger, T.A., Veith, V., Bloem, R.: Handbook of Model Checking. Springer, Cham (2018). https://doi.org/10.1007/978-3-319-10575-8
11. Danos, V., Laneve, C.: Formal molecular biology. Theoret. Comput. Sci. **325**(1), 69–110 (2004)
12. de Maria, E., Despeyroux, J., Felty, A.P.: A logical framework for systems biology. In: Fages, F., Piazza, C. (eds.) FMMB 2014. LNCS, vol. 8738, pp. 136–155. Springer, Cham (2014). https://doi.org/10.1007/978-3-319-10398-3_10
13. Despeyroux, J., Chaudhuri, K.: A hybrid linear logic for constrained transition systems. In: Post-Proceedings of TYPES 2013. Leibniz International Proceedings in Informatics, vol. 26, pp. 150–168. Schloss Dagstuhl-Leibniz-Zentrum fuer Informatik (2014)
14. Despeyroux, J., Olarte, C., Pimentel, E.: Hybrid and subexponential linear logics. Electron. Notes Theor. Comput. Sci. **332**, 95–111 (2017)
15. Gregorio, A.D., Bowling, S., Rodriguez, T.A.: Cell competition and its role in the regulation of cell fitness from development to cancer. Dev. Cell **38**(6), 621–634 (2016)
16. Ding, J., Trippa, L., Zhong, X., Parmigiani, G.: Hierarchical Bayesian analysis of somatic mutation data in cancer. Ann. Appl. Stat. **7**(2), 883–903 (2013)
17. Enderling, H., Chaplain, M.A., Anderson, A.R., Vaidya, J.S.: A mathematical model of breast cancer development, local treatment and recurrence. J. Theor. Biol. **246**(2), 245–259 (2007)
18. Fages, F., Soliman, S., Chabrier-Rivier, N.: Modelling and querying interaction networks in the biochemical abstract machine BIOCHAM. J. Biolog. Phys. Chem. **4**(2), 64–73 (2004)
19. Gavaghan, D., Brady, J.M., Behrenbruch, C., Highnam, R., Maini, P.: Breast cancer: Modelling and detection. Comput. Math. Methods Med. **4**(1), 3–20 (2002)
20. Girard, J.Y.: Linear logic. Theor. Comput. Sci. **50**, 1–102 (1987)
21. Hofestädt, R., Thelen, S.: Quantitative modeling of biochemical networks. In: Silico Biology, vol. 1, pp. 39–53. IOS Press (1998)
22. Iuliano, A., Occhipinti, A., Angelini, C., Feis, I.D., Lió, P.: Cancermarkers selection using network-based Cox regression: a methodological andcomputational practice. Front. Physiol. **7** (2016)
23. Iuliano, A., Occhipinti, A., Angelini, C., Feis, I.D., Liò, P.: Combining pathway identification and breast cancer survival prediction viascreening-network methods. Front. Genet. **9** (2018)
24. de Jong, H., Gouzé, J.L., Hernandez, C., Page, M., Sari, T., Geiselmann, J.: Qualitative simulation of genetic regulatory networks using piecewise-linear models. Bull. Math. Biol. **66**(2), 301–340 (2004)
25. Kingman, J.F.C.: Poisson Processes, Oxford Studies in Probability, vol. 3. The Clarendon Press, Oxford University Press, New York (1993). Oxford Science Publications
26. Knutsdottir, H., Palsson, E., Edelstein-Keshet, L.: Mathematical model of macrophage-facilitated breast cancer cells invasion. J. Theor. Biol. (2014)

27. Mushlin, S.B., Greene, H.L.: Decision Making in Medicine: An Algorithmic Approach, 3e (Clinical Decision Making Series), 3rd edn. (2009)
28. Olarte, C., Chiarugi, D., Falaschi, M., Hermith, D.: A proof theoretic view of spatial and temporal dependencies in biochemical systems. Theor. Comput. Sci. **641**, 25–42 (2016)
29. Phillips, A., Cardelli, L.: A correct abstract machine for the stochastic pi-calculus. In: BioConcur: Workshop on Concurrent Models in Molecular Biology (2004)
30. Regev, A., Panina, E.M., Silverman, W., Cardelli, L., Shapiro, E.: BioAmbients: an abstraction for biological compartments. Theor. Comput. Sci. **325**(1), 141–167 (2004)
31. Regev, A., Silverman, W., Shapiro, E.Y.: Representation and simulation of biochemical processes using the π-calculus process algebra. In: Proceedings of the 6th Pacific Symposium on Biocomputing, pp. 459–470 (2001)
32. Rogers, Z.N., et al.: A quantitative and multiplexed approach to uncover the fitness landscape of tumor suppression in vivo. Nat. Methods **14**(7), 737–742 (2017)
33. Savage, N.: Computing cancer software models of complex tissues and disease are yielding a better understanding of cancer and suggesting potential treatments. Nature **491**, s62–s63 (2012)
34. Shihab, H.A., et al.: An integrative approach to predicting the functional effects of non-coding and coding sequence variation. Bioinformatics **31**(10), 1536–1543 (2015)
35. Talcott, C., Dill, D.: Multiple representations of biological processes. Trans. Comput. Syst. Biol. 221–245 (2006)
36. Venkataram, S., et al.: Development of a comprehensive genotype-to-fitness map of adaptation-driving mutations in yeast. Cell **166**(6), 1585–1596.e22 (2016)
37. Wynn, M.L., Consul, N., Merajver, S.D., Schnell, S.: Logic-based models in systems biology: a predictive and parameter-free network analysis method. Integr. Biol. **4**(11), 1323 (2012)
38. Xavier, B., Olarte, C., Reis, G., Nigam, V.: Mechanizing linear logic in Coq. Electr. Notes Theor. Comput. Sci. **338**, 219–236 (2018)

Random Chromatin Neighborhoods in 2n=40 *Mus m. domesticus* Meiotic Cells: P-Percolation and Image Segmentation

Soledad Berríos[1], Julio López Fenner[2(✉)], and Aude Maignan[3(✉)]

[1] Programa Genética Humana, ICBM, Facultad de Medicina,
Universidad de Chile, Santiago, Chile
sberrios@med.uchile.cl

[2] Ingeniería Matemática, Facultad de Ingeniería y Ciencias,
Universidad de La Frontera, Temuco, Chile
julio.lopez@ufrontera.cl

[3] Univ. Grenoble Alpes, CNRS, Grenoble INP, LJK, 38000 Grenoble, France
aude.maignan@univ-grenoble-alpes.fr

Abstract. Dynamical organization of bivalents during meiosis may play a significant role in the appearance of joint genic expression, as the result of persistent overlapping domains of constitutive pericentromeric heterochromatin coming from different bivalents during pachytene.

We present early findings of an ongoing research in which we apply standard global threshold segmentation techniques for assessing correlation between observed sizes of sets of overlapping heterochromatin domains in spreads of meiotic *Mus m. domesticus* 2n=40 spermatocytes during pachytene and their simulated counterparts.

Simulated spreads were produced *ad libitum* using a special non homogeneous Bernoulli site percolation process, called P-Percolation, acting upon the fullerene's dual C'_{1200}, in which clusters of pericentromeric heterochromatin are depicted by random chromatin neigborhoods (ranches) arising from a process characterized by a vector P of independent Bernoulli random variables. We show that under the hypothesis made, a vector P of at least three dimensions is needed for establishing coherence between observed and simulated values.

Keywords: Random chromatin neighborhoods · Pachytene ·
2n=40 *Mus m. domesticus* · Inhomogeneous Bernoulli site percolation ·
Fullerene · Image segmentation

1 Introduction

The mouse *Mus m. domesticus* 2n=40 has been extensively used for modeling purposes for a number of remarkable characteristics. Among them, the morphol-

Supported by LabEx PERSYVAL-Lab (ANR-11-LABX-0025-01) funded by the French program *Investissement d'avenir*.

M. Chaves and M. A. Martins (Eds.): MLCSB 2018, LNCS 11415, pp. 142–156, 2019.
https://doi.org/10.1007/978-3-030-19432-1_9

ogy characteristics of having practically indistinguishable chromosomes (save for the sex chromosomes X and Y) makes it particularly well suited for attempting probabilistic descriptions in the dynamics of chromosome's associations. In [2] and [4] a specific stage of meiosis - pachytene - has been considered for establishing a probabilistic model of associations of bivalents, in terms of simple statistics of the observed clusters appearing in preparations (prophase meiotic nuclei), called spreads or squashes. In them, pericentromeric heterochromatin surrounding the bivalent's short arm synaptonemal complexes is highlighted, revealing domains of overlapping heterochromatin, also called association domains.

As stated in [1], these regions allow - for example - joint expression of genes coming from different bivalents, hence the associations themselves could be conceived as a form of dynamical organization from which joint genic expression can be produced or expressed.

The modeling process considered first the most eye-catching features of the observed phenomena, namely the probabilistic description of heterochromatin clusters of the telomere's bivalents attached to the nuclear envelope during pachytene.

First statistics for these clusters were established by considering the number of bivalents in each subcluster, which lead to considering partitions of the number 19 (since for *Mus m. domesticus* 2n=40 there are 19 autosomal bivalents plus the sex bivalent XY in pachytene state) by observing a set of 400 spreads, reported in [2], in which bivalents were represented by indistinguishable objects being randomly placed at the nuclear envelope. The envelope was assumed to be planar and with an hexagonal tiling.

Then, in [4] the nuclear envelope was replaced by a locally planar six-regular graph, which means that each position upon the nuclear envelope accepts exactly six neighbors. Finally, in [1] a clustering process was proposed based on the notion of *Random chromatin neighborhoods* (ranchs) and an association process derived from a special type of site non-homogeneous percolation processes upon vertices of a fullerene's dual C'_{1200}, called P-percolation.

Hence, for the purposes of this work, we simulate spreads by means of random chromatin neighborhoods generated by a P-percolation process, as in [1], that represent both the pericentromeric heterochromatin (CPCH) surrounding the Synaptonemal Complexes (SC's) and the SC's themselves, attached to the envelope, which is modelled by C'_{1200}.

Since ranches uniformly distributed in a bounded domain will overlap and hence produce clusters, their size distribution will be determined in a fashion depending solely upon the Bernoulli random variables driving the percolation process.

In this work we continue the interrogation of the model by processing the original data in order to estimate coherence between our theoretical model and the biological subject, by relating the observed nuclear surface being taken by the ranches, on one side, and the simulated surface being taken by the clusters in dependency upon the individual Bernoulli random variables used to construct the associated percolation process, on the other.

The article proceeds then as follows: In the next section, Materials and Methods, we present the data acquisition method used for obtaining the statistics of the cluster's distribution and explain the main biological concepts used throughout this article. We further explain the global threshold segmentation method used for estimating observed chromatin surfaces in the spreads.

We proceed then with the model of the nuclear envelope and explain the terminology used, in particular P-Percolation and the notion of ranches. We provide some theoretical results for the average chromatin surface in terms of the underlying Bernoulli random variables that drive the percolation.

We then briefly discuss the determination of length 3 for the P-vector in the percolation as yielding optimal coherence between simulated and observed values. We close the article with a brief discussion of our results as well as providing some possible lines of future research.

2   Materials and Methods

A dataset of 400 photographs (in jpg format) was considered, which are the result of removing the nuclear envelope of spermatocytes in the meiosis's pachytene state, exposing DNA related structures associated to the individual bivalents: The synaptonemal complex and the constitutive pericentromeric heterochromatin. Each one of these preparations are called spreads.

2.1   Biology and Data

The methodology of the data collection is described through 4 steps:

- **Animals**: Spermatocytes were taken from two male three-month-old *Mus m. domesticus* 2n=40 C3H mice. Mice were maintained at 22 °C with a light/dark cycle of 12/12 h and fed *ad libitum*. Procedures involving the use of - and upon - mice were approved by the Animal Ethics Committee of the Faculty of Medicine, Universidad de Chile.
- **Spermatocyte nuclear spreads**: Spermatocyte spreads were obtained following the procedure described by Peters et al. [7], and Page et al. [6]. Briefly, a testicular cell suspension in 100 mM sucrose was spread onto a slide dipped in 1% para-formaldehyde in distilled water containing 0.15% Triton X-100 then left to dry for two hours in a moist chamber. The slides were subsequently washed with 0.08% Photoflow (Kodak), air-dried, and re-hydrated in PBS.
- **Immunochemical identification of bivalents**: The slides were incubated for 24 h at 4 °C in a moist chamber with the primary antibodies: mouse anti-SYCP3 1:100 (Santa Cruz, 74569) and rabbit anti-H3K9me3 1:200 (Abcam ab8580). Then, the slides were incubated for 30 min at room temperature with the secondary antibodies: FITC-conjugated goat anti-mouse IgG (1:50) (Sigma), or Texas red-conjugated goat anti-rabbit IgG (1:200) (Jackson). Finally, slides were rinsed in PBS and mounted in Vectashield (Vector).

- **Image registration:** Stained slides were viewed and photographed on a Nikon Optiphot or an Olympus BX61 epifluorescent microscope at 1000X equipped with Nikon PL-APO 100X, 1.30 NA objective lenses to monitor staining and assess results. Individual photographs were produced with a DS camera control unit DS-L1 Nikon or captured with an Olympus DP70, with digital center-weighted average metering mode and manual exposure, and with X and Y Resolution set to 0.01 in RGB format (JPG) of 2560×1920 pixels.

A bivalent is a complex structure consisting in two synapsed homologous chromosomes through the synaptonemal complex SC. All SC's appear green in the photographs and each is surrounded by a cloud of constitutive pericentro-metric heterochromatin, CPCH. If two bivalents exhibit overlapping domains of CPCH near to their short arms (tainted red), then they are said to be in asso-ciation. Hence in the spreads, association domains appear as a connected set of pixels tainted red and the number of bivalents in association correspond to the number of green tainted SC's belonging to a given connected domain.

We count the number or bivalents in each connected domain and assign to the spread the corresponding *Partition of the number 19* thus determined.

As an illustration of this procedure, refer to Fig. 1, which depicts a represen-tative instance of a pachytene nuclear spread. Observe the 19 indistinguishable autosomal bivalents and the sex chromosomes X and Y. The partition of 19 associated to this photograph corresponds to $19 = 7 + 4 + 2 + 1 + 1 + \dots + 1$, which means that the biggest cluster contains 7 bivalents, the second contains 4, etc.

We say that the spread in Fig. 1 belongs to *1st-Class* 7 and *2nd-Class* 4.

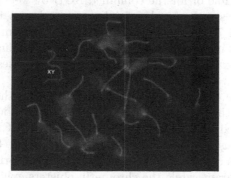

Fig. 1. A representative pachytene spermatocyte spread of $2n = 40$ *Mus domesti-cus* mice treated by immunocytochemical techniques. In red the pericentromeric hete-rochromatin domains, in green the synaptonemal complexes. The (sex) bivalent XY is indicated. (Color figure online)

All 400 spread photographs were classified and the distribution made avail-able in the previous works, see [1,2] and [4]. For the purposes of this article, we denote by $f_O^1[i]$ the frequence of spreads belonging to the 1st – Class i. Similarly, we denote by $f_O^2[i]$ the frequence (in %) of spreads belonging to the 2nd – Class i.

The following table, reprinted partially from [1], gives the 1st-Class frequencies (Table 1):

Table 1. Observed 1st-Class frequencies of sizes 1 to 19

i	1	2	3	4	5	6	7	8	9	10	11	12	$\cdots$
$f_O^1[i]$	0.00	0.50	11.25	23.25	23.50	18.50	10.25	6.50	4.25	1.25	0.75	0.00	$\cdots$

Analogously, we can consider the second Class (meaning the frequencies of the second biggest cluster) and obtain the table for the f_O^2 values. This values were also reported in [1] and hence we omit them here. But for establishing a quality parameter for our approximations, we will refer to the quadratic error in the determination of the associated 1st Class (resp. 2d-Class) frequencies, which we call ϵ_1^2 (resp. ϵ_2^2), see Eq. (4.1) below.

2.2 Global Threshold Image Segmentation for Chromatin Surface Determination

Images in RGB format are presented in three frames, one for each color, and correspond numerically to a matrix of pixels in 3D dimensions. Pixel values in a given frame are either integers between 0 and 256 or normalized between 0 and 1, we used only normalized images in our analysis. A Grayscale image can be produced for each frame or for the combined RGB by weighting the individual frames according to standard polynomial conversions, as in luminosity (21% Red, 72% Green and 7% Blue), which follows ITU-R recomendation BT709 [8] (Fig. 2). This procedure has been used elsewhere for luminance derived image processing, see for example [3]. An histogram of intensities was produced for each image, as in Fig. 3.

Background noise can be cut off in order to produce the nucleus including the chromatin, and the chromatin, respectively. Distinct outliers (in our case, zones with illuminated pixels that do not correspond to CPCH) from non related structures in the image can be removed by hand. In Fig. 3, background noise is contained in the interval $[0, 0.2]$ approximately, which correspond to position $k_l = 17$ in the histogram, while the chromatin clusters exhibit pixel intensities above 0.4 (or position $k_u \geq 27$). Figure 4 shows the nucleus and chromatin for the example image where the cut-off pixel intensities were set to be 0.25 and 0.475. With these values, the relative surface being taken by the chromatin, which is proportional to the number of highlighted pixels, corresponds to an equivalent of 27.9 % of the total surface of the nucleus. In terms of the graph with 602 nodes, this corresponds to 168 nodes, which is consistent with a P-percolation generated by a P-vector of probabilities with at least three dimensions, see discussion below.

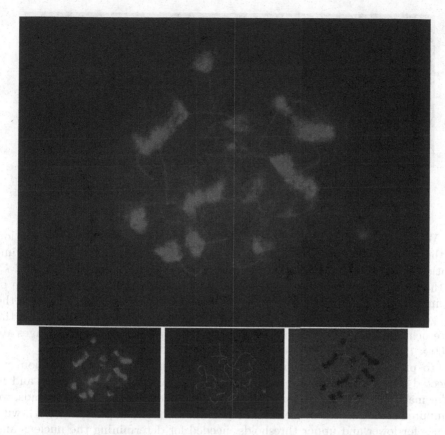

Fig. 2. Image RAD50001 in (a) composite (R, G, B) representation, (b) Red (c) Green and (d) Blue (in Gray scale for better visualization) (Color figure online)

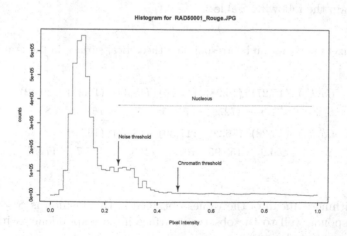

Fig. 3. Histogram for Red frame of Image RAD50001. Noise threshold correspond to $k_l = 17$, chromatin threshold to $k_u \geq 27$ (Color figure online)

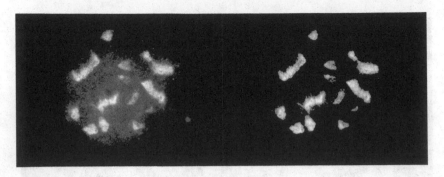

Fig. 4. Nucleus and chromatin content for image RAD50001

We use a gray-level histogram based thresholding for assessing segmentation of the images, as depicted in [9], in which a global threshold is used for discrimination of the structures. We use two levels of intensities at positions k_l and k_u of the histogram, considered as a vector. Hence, if H denotes the histogram of an image, and $f(m,n)$ denotes the pixel intensity at coordinates (m,n) of the image, the nuclei will be the set of pixels that satisfy $H(k_l) > f(m,n)$ and the heterochromatin cluster will satisfy $H(k_u) > f(m,n)$. In our case, all images are 2560×1920 matrices with pixel intensities in $[0,1]$.

To provide the best mean chromatin surface in a first approximation, we selected 20 spreads satisfying homogeneity criteria: similar in intensity and in color made by the same person with the same process. For those 20 spreads, we computed the mean of surface chromatin S for different couples of (k_l, k_u), with values for lower and upper thresholds, needed for determining the nucleus and the chromatin clusters, satisfying the restrictions: $15 \leq k_l \leq 20$, $27 \leq k_u \leq 33$ and $12 \leq k_u - k_l \leq 13$.

We obtain the following Table 2:

Table 2. Pixel thresholds and corresponding theoretical surface in the graph approximation

(k_l, k_u)	(15,27)	(16,28)	(17,29)	(18,30)	(19,31)	(20,32)
S	171.5	172.2	174.75	177	181	186

(k_l, k_u)	(15,28)	(16,29)	(17,30)	(18,31)	(19,32)	(20,33)
S	156.9	158.95	161	163.45	167.7	172.7

The arithmetic mean of the values obtained in this fashion is $\hat{S} = 170.2625$. This corresponds well to the observed surface in correspondence with the 602 size of the associated graph.

Since the contours of the heterochromatin (CPCH clusters) and of the nucleus itself are not sharp enough, because the nuclear envelope was removed before taking the photograph, it is hard to select the best objective parameter couple (k_l, k_u) based on pixel gradients alone or from first principles. Other approaches like Sobel segmentation as in [5], Renyi's entropy [9] or even a cellular neural networks (CNN) based algorithm [10] are currently under review and will not be reported here.

3   P-Percolation and Ranches

For self-containedness purposes we recall briefly the model of the clustering process based on Random chromatin neighborhoods (ranches) [1], which are the result of a special site non-homogeneous percolation processes acting upon vertices of a fullerene's dual C'_{1200}.

Recall that the dual of the fullerene C_{1200} is a graph of 602 nodes, all but 12 of them are six-regular (those exceptional 12 are five-regular), which means that we can safely consider the entire graph as being six-regular, a condition required by first principles in [2] that has been preserved in [1] and [4].

The P-Percolation process upon the six-regular graph proceeds then as follows: a random vertex is chosen, which will be the position of the Synaptonemal Complex SC of a given bivalent upon the nuclear envelope. The percolation process now selects vertices in an outward direction, layer by layer according to independent Bernoulli random variables with probabilities P_1, P_2, etc.

Each Bernoulli process $B(P_i)$ acts upon vertices in the i-th layer, selecting those sites that are reachable from the preceding layer (i.e., bonds between vertices of the same layer are not relevant and thus not considered). Reachable means here that an already selected vertex in the previous layer exists that lies at distance one of the candidate vertex (Fig. 5).

Hence, this site percolation process is a special type in which the set of selected vertices in a neighborhood of up to κ layers represent a random chromatin neighborhood centered at the position of the starting SC, which we call *ranch*. This process is repeated according to the number of SC's present in the spermatocyte and the induced clustering is determined by overlapping vertices chosen from ranches belonging to different SC's and fully determined by the vector of probabilities $P = (P_1, P_2, \ldots, P_\kappa)$.

As a consequence, let us denote by MC the set of vertices of $G = C'_{1200}$ that represent the position of the $SC's$ and by $ranch^P(w)$ the ranch associated to each $w \in MC$, (notice that by definition $w \in ranch^P(w)$, for all $w \in MC$). In Fig. 6, an element of MC is represented by a big dot. An element of a ranch is represented by a small dot. When two ranches overlaps they belongs to the same cluster. All the elements belonging to the same cluster have the same color.

Let S be the theoretical chromatin surface: $S = | \cup_{v \in MC} ranch^P(v)|$. The expected (mean) chromatin surface for a 3-dimensional P-percolation process is given by

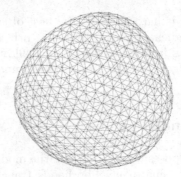

Fig. 5. The discrete nuclear envelope model $G = C'_{1200}$.

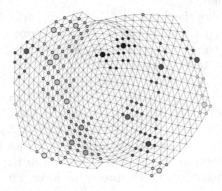

Fig. 6. A spread simulation with $P = (P_1, P_2, P_3) = (0.6, 0.415, 0.3)$. The cluster description is $19 = 6 + 3 + 2 + 2 + 1 + 1 + 1 + 1 + 1 + 1$

Theorem 1. *Let* $P = (P_1, P_2, P_3)$ *be a probability vector of length* $\kappa = 3$ *for the* P-*percolation upon* $G = C'_{1200}$. *Then*

$$\mu(S) = 19 + 583 \sum_{k=1}^{18} (-1)^{k+1} q_{k1} + 602 q_{19,1}$$

with

$$q_{k1} = \frac{19^k}{602^k} P_1^k \sum_{\substack{n_1 + n_{2a} + n_{2b} + n_{3a} + n_{3b} = k \\ n_1 \leq 6,\, n_{2a} \leq 6\, n_{2b} \leq 6 \\ n_{3a} \leq 6,\, n_{3b} \leq 12}} \binom{n_1}{6} \binom{n_{2a}}{6} \binom{n_{2b}}{6} \binom{n_{3a}}{6} \binom{n_{3b}}{12}$$

$$\times P_2^{k-n_1} (2 - P_1)^{n_{2b}} P_3^{n_{3a} + n_{3b}} (3 - P_1 - P_1 P_2 (2 - P_1))^{n_{3b}}$$

Proof: Consider a vertex v of $G = C'_{1200}$ and denote by q_k the probability that v belongs to – at least – k ranches. Put

$$q_k = q_{k0} + q_{k1}$$

in which q_{k0} is the probability of v to belong to at least k ranches knowing that $v \in MC$ and q_{k1} is the probability of v to belong to at least k ranches knowing that $v \notin MC$.

Since the number of SC is 19 and the number of vertices of G is 602,

$$\mu(S) = 602 \sum_{k=1}^{19} (-1)^{k+1} q_k$$

The probability for a vertex to be in MC is $\frac{19}{602}$ and $v \in ranch^P(v)$, thus $q_{10} = \frac{19}{602}$ and $q_{k0} = \frac{19}{602} q_{k-1,1}$ for $k > 1$.

$$\text{We obtain } \mu(S) = 602 \sum_{k=1}^{19} (-1)^{k+1} q_k$$

$$= 602 \sum_{k=1}^{19} (-1)^{k+1} (q_{k0} + q_{k1})$$

$$= 602 \left(q_{10} + \sum_{k=1}^{18} (-1)^{k+1} \left(q_{k1} - \frac{19}{602} q_{k1} \right) + q_{19,1} \right)$$

$$= 602 \left(q_{10} + \sum_{k=1}^{18} (-1)^{k+1} \frac{583}{602} q_{k1} + q_{19,1} \right).$$

$$= 19 + 583 \sum_{k=1}^{18} (-1)^{k+1} q_{k1} + 602 q_{19,1}$$

Notice that q_{k1} depends on κ and P.

In the following we restrict ourselves to $\kappa = 3$, i.e., $P = (P_1, P_2, P_3)$. A generalization of the formula for $\kappa > 3$ can be readily obtained and will be omitted here.

Now assume that v is at the center of a local view of the graph, see Fig. 7:

- If $w \in MC$ and $d(v, w) = 1$, the vertex v belongs to $ranch^P(w)$ with probability $\delta_1 = P_1$.
- If $w \in MC$ and $d(v, w) = 2$, there are two possibilities:
  - $w = Z(2, 2j)$ with $j = 0, 1, 2, 3, 4$ or 5, then v belongs to $ranch^P(w)$ with probability $\delta_{2a} = P_1 P_2$
  - if $w = Z(2, 2j + 1)$ with $j = 0, 1, 2, 3, 4$ or 5, the vertex v belongs to $ranch^P(w)$ with probability $\delta_{2b} = P_1 P_2 (2 - P_1)$
- If $w \in MC$ and $d(v, w) = 3$, there are also two cases:
  - if $w = Z(3, 3j)$ with $j = 0, 1, 2, 3, 4$ or 5, the vertex v belongs to $ranch^P(w)$ with probability $\delta_{3a} = P_1 P_2 P_3$
  - if $w = Z(3, 3j + 1)$ or $w = Z(3, 3j + 2)$ with $j = 0, 1, 2, 3, 4$ or 5, the vertex v belongs to $ranch^P(w)$ with probability $\delta_{3b} = P_1 P_2 P_3 (3 - P_1 - P_1 P_2 (2 - P_1))$

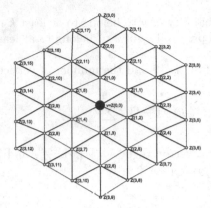

Fig. 7. Reachable set at distance 3 for the (site) Bernoulli P-Percolation process (reproduced from [1]).

Let us fix a quintuplet of integers $(n_1, n_{2a}, n_{2b}, n_{3a}, n_{3b})$ such that $n_1 \leq 6$, $n_{2a} \leq 6$, $n_{2b} \leq 6$, $n_{3a} \leq 6$, and $n_{3b} \leq 12$ and let us assume that there are at least

- n_1 elements of MC at distance 1 of v;
- n_{2a} elements of MC at distance 2 of v which have a position $Z(2, 2j)$
- n_{2b} elements of MC at distance 2 of v which have a position $Z(2, 2j+1)$
- n_{3a} elements of MC at distance 3 of v which have a position $Z(3, 3j)$
- n_{3b} elements of MC at distance 3 of v which have a position $Z(3, 3j+1)$ or $Z(3, 3j+2)$.

With this assumption, the probability for v to belong to a set of at least $n_1 + n_{2a} + n_{2b} + n_{3a} + n_{3b} = k$ vertices is given by

$$\frac{19^k}{602^k} \binom{n_1}{6} \delta_1^{n_1} \binom{n_{2a}}{6} \delta_{2a}^{n_{2a}} \binom{n_{2b}}{6} \delta_{2b}^{n_{2b}} \binom{n_{3a}}{6} \delta_{3a}^{n_{3a}} \binom{n_{3b}}{12} \delta_{3b}^{n_{3b}}.$$

From this, we obtain

$$q_{k1} = \frac{19^k}{602^k} \sum_{\substack{n_1+n_{2a}+n_{2b}+n_{3a}+n_{3b}=k \\ n_1 \leq 6,\, n_{2a} \leq 6\, n_{2b} \leq 6 \\ n_{3a} \leq 6,\, n_{3b} \leq 12}} \binom{n_1}{6} \delta_1^{n_1} \binom{n_{2a}}{6} \delta_{2a}^{n_{2a}} \binom{n_{2b}}{6} \delta_{2b}^{n_{2b}} \binom{n_{3a}}{6} \delta_{3a}^{n_{3a}} \binom{n_{3b}}{12} \delta_{3b}^{n_{3b}}$$

$$= \frac{19^k}{602^k} P_1^k \sum_{\substack{n_1+n_{2a}+n_{2b}+n_{3a}+n_{3b}=k \\ n_1 \leq 6,\, n_{2a} \leq 6\, n_{2b} \leq 6 \\ n_{3a} \leq 6,\, n_{3b} \leq 12}} \binom{n_1}{6} \binom{n_{2a}}{6} \binom{n_{2b}}{6} \binom{n_{3a}}{6} \binom{n_{3b}}{12}$$

$$\times P_2^{k-n_1} (2 - P_1)^{n_{2b}} P_3^{n_{3a}+n_{3b}} (3 - P_1 - P_1 P_2 (2 - P_1))^{n_{3b}}$$

Remark: Since for $k \geq 6$, q_{k1} is almost negligible, we can safely truncate the series.

4 Surface, Clusters and Best P−Percolation

This section is devoted to the correlation between the theoretical surface and the non homogeneous Bernoulli site percolation process. For the present analysis, we consider a P-percolation with a P-vector of length 3 and express the chromatin's surface S in terms of the equivalent number of vertices in C'_{1200}.

Figure 8 depicts the contour lines of the theoretical chromatin surface together with the empirical best approximation obtained for 1st-Class, i.e., the biggest cluster. Best approximation means that the selected vector minimizes the sum of the squares in observed (f_O^1) versus simulated (f_S^1) 1st-Class cluster distribution (respectively 2^d-Class):

$$\epsilon_1^2 := \sum_{i=1}^{19} |f_O^1(i) - f_S^1(i)|^2 \tag{4.1}$$

In Fig. 8(a), we consider first $P_3 = 0$. The red dots correspond to those (P_1, P_2) values for which ϵ^1 is minimal under P−percolation with $P = (P_1, P_2, 0)$. The computation was made with 10^5 iterations. The black line corresponds to a spline interpolation with a degree 3 piece-wise polynomial in P_2 that approximates the red dots. Contour lines of the surface equation $S = \sigma$ (σ from 100 to 400 by 5) were added in grey in the background. We observe that a few contour lines cross the best approximation line. In the borders, the minimal (lower) and maximal (upper) contour lines have been marked in dashed blue.

The same observation applies for $P_3 = 0.2$ (Fig. 8(b)), $P_3 = 0.3$ (Fig. 8(c)) and $P_4 = 0.4$ (Fig. 8(d)).

The curve which fits best the experimental observations is obtained with $P_3 = 0.3$, with P−percolation probability vectors as given in the following Table 3:

Table 3. P-percolation vectors

P_1	P_2	P_3	ϵ_1	ϵ_2	S
0.5	0.53	0.3	3.06	9.13	170.16
0.55	0.467	0.3	3.10	8.45	170.2
0.6	0.415	0.3	2.97	7.78	170.43
0.65	0.375	0.3	2.72	7.62	171.53
0.7	0.335	0.3	2.37	7.60	171.5

The corresponding 1st-Class i frequencies are found to be (Table 4):

Figure 6 above showed a spread simulation with $P = (0.6, 0.415, 0.3)$. Moreover the simulated mean surface turns out to be 169.4656, in close agreement with the expected value.

Table 4. 1st-Class i frequencies with $P = (0.6, 0.415, 0.3)$

i	1	2	3	4	5	6	7	8	9	10
$f_o^1[i]$	0.000	0.822	10.630	22.461	22.448	16.647	11.044	6.860	3.992	2.395

i	11	12	13	14	15	16	17	18	19	
$f_o^1[i]$	1.321	0.725	0.371	0.174	0.074	0.025	0.008	0.003	0.000	

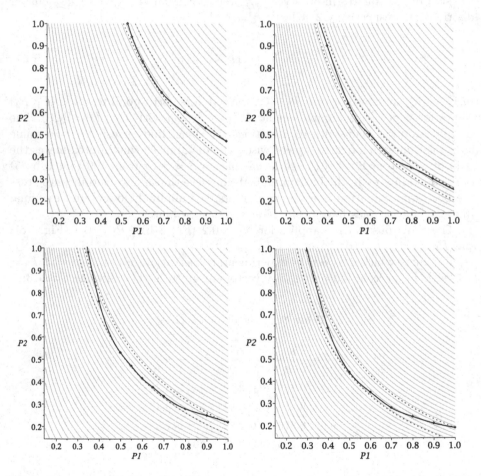

Fig. 8. Best approximation for 1st-Class obtained with P-percolation $P = (P_1, P_2, P_3)$ in (P_1, P_2)–space (in black) and contour lines of the surface equation $S = \sigma$ (σ from 100 to 400 by 5) coupled with the lower and uooer contour lines $S = \sigma_{min}$ and $S = \sigma_{max}$ (in dashed blue): (a) $P_3 = 0$, $\sigma_{min} = 187$, $\sigma_{max} = 200$, (b) $P_3 = 0.2$, $\sigma_{min} = 172$, $\sigma_{max} = 184$. (c) $P_3 = 0.3$, $\sigma_{min} = 170$, $\sigma_{max} = 181$,(d) $P_3 = 0.4$, $\sigma_{min} = 164$, $\sigma_{max} = 183$. (Color figure online)

5   Conclusion

In this work we considered P- percolation as the underlying dynamics of CPCH cluster formation in spermatocytes of *Mus m. domesticus* 2n=40 during pachytene. Using simple threshold segmentation analysis we gained insight upon the set of suitable P vectors that provide best approximations to the observations.

We cannot help but to emphasize that neither an homogeneous nor a non homogeneous percolation process of length 2 suffices to adequately explain the observed distribution of the 1st- Class cluster distribution, so that necessarily a P-percolation process with P-vector of length tree – at least – seems appropriate for a proper description of the underlying randomness in this state.

The non homogeneous kind of percolation represents also an interesting aspect to be further elucidated in future works. Moreover, even if the contour lines of the spread images are not sharp, due to the intrinsic technique of removing the envelope, but also because of the staining techniques used for revealing the CPCH's and SC's structures during pachytene, further improvements in accuracy or P length determination should be obtained through better image segmentation methods, a task that is presently under development.

From this perspective, the results show that, while a three vector approximation yields promising expectations, we need still to work also on other characteristics of the distributions, as for example the 2^{nd}Class too. Indeed, we focused here upon the quality of our model with respect to the results obtained mainly for the $1^{st}Class$ and the corresponding approximation of the surface.

We see that the results on ϵ_2 are not yet as good as those of ϵ_1. At the present time, we haven't decided whether a P vector of length greater than three can furnish an even better approximation for the distributions in terms of both ϵ_1 and ϵ_2 and whether this would provide a chromatin surface in better correlation with the observed surface. This is a matter of ongoing research.

We thank the referees for their valuable comments.

References

1. Berríos, S., López-Fenner, J., Maignan, A.: Simulating aggregates of bivalents in 2n=40 mouse meiotic spermatocytes through inhomogeneous site percolation processes. J. Math. Biol. (2018). https://doi.org/10.1007/s00285-018-1254-6
2. Berríos, S., Manterola, M., Prieto, Z., López-Fenner, J., Page, J., Fernández Donoso, R.: Model of chromosome associations in mus domesticus spermatocytes. Biol. Res. **43**(3), 275–285 (2010)
3. Liu, J., Qiu, G., Shen, L.: Luminance adaptive biomarker detection in digital pathology images. Procedia Comput. Sci. **90**, 113–118 (2016)
4. López-Fenner, J., Berríos, S., Manieu, C., Page, J., Fernández-Donoso, R.: Bivalent associations in mus domesticus 2n=40 spermatocytes are they random? Bull. Math. Biol. **76**(8), 1941–1952 (2014)
5. Morar, A., Moldoveanu, F., Gröller, E.: Image segmentation based on active contours without edges. In: 2012 IEEE 8th International Conference on Intelligent Computer Communication and Processing. pp. 213–220. IEEE (2012)

6. Page, J., Suja, J.A., Santos, J.L., Rufas, J.S.: Squash procedure for protein immunolocalization in meiotic cells. Chromosome Res. 6(8), 639–642 (1998)

7. Peters, A.H., Plug, A.W., van Vugt, M.J., De Boer, P.: Short communications a drying-down technique for the spreading of mammalian meiocytes from the male and female germline. Chromosome Res. 5(1), 66–68 (1997)

8. Recommendation, I.: 709–5, parameter values for the hdtv standards for production and international programme exchange. ITU Radiocommunication (2002)

9. Sahoo, P., Wilkins, C., Yeager, J.: Threshold selection using Renyi's entropy. Pattern Recogn. 30(1), 71–84 (1997)

10. Vilariño, D.L., Cabello, D., Pardo, X.M., Brea, V.M.: Cellular neural networks and active contours: a tool for image segmentation. Image Vis. Comput. 21(2), 189–204 (2003)

Author Index

Printed in the United States
By Bookmasters

Printed in the United States
By Bookmasters